PASS YOUR AMATEUR RADIO TECHNICIAN CLASS TEST – THE EASY WAY
2018-2022 Edition

By: Craig E. "Buck," K4IA

ABOUT THE AUTHOR: "Buck," as he is known on the air, was first licensed in the mid-sixties as a young teenager. Today, he holds an Amateur Extra Class Radio License.

Buck is an active instructor and a Volunteer Examiner. The Rappahannock Valley Amateur Radio Club named him Elmer (Trainer) of the Year three times and Buck has successfully led many students through this material.

Email: K4IA@EasyWayHamBooks.com

Published by EasyWayHamBooks.com
130 Caroline St. Fredericksburg, Virginia 22401

"Easy Way" Books by Craig Buck are available at Ham Radio Outlet stores and Amazon:
"Pass Your Amateur Radio General Class Test"
"Pass Your Amateur Radio Extra Class Test"
"How to Chase, Work & Confirm DX"
"How to Get on HF"
"Prepper Communications"
"Pass Your GROL Test"

Copyright ©2018, Craig E. Buck   All Rights Reserved. No part of this material may be reproduced, transmitted or stored in any manner, in any form or by any means without the express written permission of the author. 1k

ISBN 978-1985125643

PASS YOUR AMATEUR RADIO TECHNICIAN CLASS TEST – THE EASY WAY

TABLE OF CONTENTS

> If you are stuck on a question or concept, visit my Facebook group "Ham Radio Exams." You will find plenty of folks happy to assist and encourage.

INTRODUCTION

There are many books to help you study for the amateur radio exams. Most focus on taking you through all the questions and possible answers on the multiple-choice test. The problem with that approach is that you must read three incorrect answers for every one correct answer. That's 1,272 wrong answers and 424 right answers. No wonder people get overwhelmed.

This book is different. There are no wrong answers. I'm going to take you on a journey of discovery. In the process of relating my journey through amateur radio, I'll answer every question on the Amateur Radio Technician Class exam in a way that will help you understand the answers and be entertained at the same time. **The test questions and answers are in bold print to help you focus.**

I might repeat a question because it fits into the narrative in more than one place. That will just reinforce the answer in your mind.

The second part of this book is a Quick Summary – just the question and the right answer. I've condensed what others say in over 200 pages down to 40 pages because you do not see the wrong answers.

Once you get your Technician license, you'll want to move up with my other books: "How to Pass Your Amateur Radio General Class Test – The Easy Way" and "How to Get on HF – The Easy Way." Order them today, so you'll be ready! EasyWayHamBooks.com

THE TEST

The test is 35 questions from a possible pool of 424. The questions and the answers on your test are word-for-word out of the published question pool. You know what will be on the test. The answer's order is scrambled, so answer "B" in the question pool is not necessarily "B" on your test. Don't try memorizing the multiple-choice letters.

The pool is large but only about one in a dozen questions will show up on your exam, one from each of the test subjects. I can guarantee you won't get 35 questions on Ohm's Law. Many questions ask for the same information in a slightly different format, so there aren't 424 unique questions. You need to answer 26 correctly.

There are **three license classes: Technician, General, and Amateur Extra**[1] each conferring more privileges. How hard is the Technician test? It is not hard. There is no heavy-duty algebra, calculus or trigonometry. Don't let the material intimidate you. There is no Morse code required for any class of Amateur license. Morse code is still very much alive and well on the amateur bands, and there are many reasons you might want to learn it in the future. Just don't worry about it for now.

My high-school chum, Billy, and I studied for our test by reading textbooks. We didn't know what questions would be on the test. Now, the questions and answers are all available so you can study with confidence that you are focusing on the right material.

I have personally taken students who didn't know anything about electronics, radio or math through this

---

[1] There's your first question and answer.

THE TEST

material, and they passed.   You want to learn it all but, if there's a question you just can't get, chances are it won't be on your test. And, you can miss nine out of the thirty-five questions and still pass.

Do you know what they call the medical student who graduates last in his class? "Doctor" – the same as the guy who graduated first in the class. My point is: all you need is to pass, and no one will know the difference. To pass, answer 26 out of 35 correctly.

The questions are multiple-choice, so you don't have to know the answer, you just have to recognize it. That is a tremendous advantage for the test-taker. However, it's difficult to study the multiple-choice format because you can get bogged down and confused by seeing three wrong answers for every correct one. I think that makes it harder to recognize the right answer when you see it.

The best way to study for a multiple-choice exam is to concentrate on the correct answers. Find a word or phrase you can remember. When you take the test, the correct answer should jump right out at you.

If you don't know the answer, eliminate the obviously wrong answers. Then make an educated guess. There is no penalty for guessing wrong, and you greatly increase your odds by eliminating the bogus responses.

Long ago, when my friend, Billy, and I sat for our General test, we took a bus to the FCC Building in Washington, DC. There, we met the sternest proctor I have ever suffered under. Picture the short-sleeve white shirt, skinny black tie, pocket protector, slide-rule-on-the-belt, thick eyeglasses, cigarette smoke and a full ashtray. I felt that if I did not pass, I was going straight to jail. I was sure the FBI told him I had listened to Radio Moscow.

Today, it is all much easier. Volunteer Examiners organized under the authority of the FCC, administer Amateur Radio testing. VEs are local hams who devote their time to help you get your license. In most areas, you can find a VE group testing at least once a month. There are 14 accredited VE organizations, but ARRL and W5YI are the biggest. Look on the licensing tab of ARRL.org or W5YI.org to find a test site.

On test day, you should bring a picture ID, your Social Security Number (not the card, just the number) or your FRN number,[2] a pen and pencil, and about $15. You need to ask ahead to get the exact amount and see if they prefer cash or check. The modest exam fee reimburses the VE's costs. No one profits except you. If you bring a calculator, you must clear the memories. Turn off your cell phone and don't look at it during the exam.

You will complete a simple application (Form 605), and might save a few minutes by downloading the form from the Federal Communications Commission (FCC) website and filling it out ahead of time. Your Volunteer Examiner (VE) team will select a test booklet for you. Make sure you ask them for the easy one – that's good for a cheap laugh. They have no idea which questions are in your booklet. You also get a multiple-choice, fill-in-the-circle answer sheet. The answer sheet has room for 50 questions, but you stop at 35. You can take notes and calculate on the back of the answer sheet – not in the booklet. VEs re-use the test books.

---

[2] Many VEs request a Federal Registration Number (FRN) from the FCC instead of your Social Security Number. The FRN is used only for your communications with the FCC. Search "FRN FCC" to get the link to the FCC website where you can register. Then, bring your FRN number to the test session.

THE TEST

The VE team will grade your exam while you wait and give you a pass/fail result. Don't ask them to go over the questions or tell you what you missed. They don't know and don't have time to look. When you pass, there is a bit more paperwork, and you walk out with a CSCE – a Certificate of Successful Completion of Examination. The team processes your Form 605 and CSCE, and your call sign appears in the FCC (Federal Communications Commission) database in a few days. Congratulations.

You can start transmitting when you see your call sign appear online in the FCC database. This takes about a week. **Your proof of possession is that your license appears in the FCC ULS consolidated database.**

Your license is good for ten years and you can renew without another test.

There is a two-year grace period during which you can renew your license without another test. You may not operate a transmitter during the grace period because you are not licensed.

Once you get your license, you want to make sure to keep the FCC updated on any address changes. **When correspondence from the FCC is returned as undeliverable because the grantee failed to provide and maintain a correct mailing address, revocation of the station license or suspension of the operator license may be the result.**

And while we're on the subject of the FCC, you should know that **the FCC has the power to inspect your station and records at any time**. It is highly unlikely that you would ever get a visit from the FCC, but if it happens, you need to cooperate.

HOW TO STUDY

You ace a multiple-choice exam by learning to recognize the right answer and eliminate wrong answers. Poring over the multiple-choice questions has been the traditional approach taken by most classes and license manuals. The problem with that method is you have to read through three wrong answers to every question. That is both frustrating and confusing. Why study the WRONG answers?

Here, you never see the wrong answers so the right answer should pop out on the test. You don't need to memorize the whole answer – just enough of it to recognize. Even if you don't understand a word of the question or answer, you can recognize the right answer.

I don't recommend you take practice exams until the week before the test and <u>after</u> you feel you have mastered the material. The reason I say to wait is so you don't get confused by seeing the wrong answers. You want to recognize the right answers first. The practice tests build your confidence. They don't help you learn.

Your amateur radio license is called a "ticket." You can't start a journey without one. Your Ham ticket can take you where you never thought you could go. Get your ticket first. The journey of discovery will last a lifetime.

Throughout the material, the questions and answers are in **bold print**. Hints, notes, and cheats are in *italic*. Cheats are tricks to help you remember an answer.

HOW I GOT STARTED

Begin at the beginning. I grew up in the days before personal computers, the internet, video games and 500 channels of color TV. We made our entertainment. I remember building an airplane cockpit using shoe boxes with toothpicks for control levers. We shot marbles and built elaborate obstacle courses that would make a golf course designer proud. If ants got in the way, we dumped lighter fluid down their hole and lit it. No mercy. That was about third grade.

These were the days when you rode a bike without a helmet; cars didn't have seatbelts, and you stayed out until the street lights came on. We played in the woods, caught snakes and lit cherry bombs. Model boats and planes filled my shelves. In fifth-grade science class, we built a crystal radio set. It only tuned one station. At night, I would tuck the earpiece under my pillow and fall asleep listening to basketball games. Basketball on the radio took a lot of imagination and held no interest for me. I was just amazed to hear radio signals coming from a radio made with a rock and no batteries. That station was about 30 miles away.

Tuning the family's Crosley 5-tube AM radio, I discovered "skip" – the bouncing of radio waves off the ionosphere. **Radio waves are electromagnetic. The ionosphere is the part of the atmosphere that enables propagation around the world.** Radio waves bounce off the ionosphere and come down a distance away. It is possible to hear stations from hundreds or thousands of miles off as you may have discovered with your car's AM radio at night.

Around 7th grade, I got a shortwave radio kit for Christmas. Building it, I learned **an electrical wiring diagram that uses standard component symbols**

is called a schematic. *Hint: The short answer is, "A wiring diagram is a schematic."* **Electrical schematics represent the way components are interconnected. The schematic symbols represent electrical components.**

I burned my fingers many times while learning to solder. **We use rosin-core solder** because Dad's acid-core plumbing solder would corrode the joint. I also learned that cold solder joints are not good. **Cold solder joints look grainy and dull** and make a bad connection. You need to reheat the joint and try again. Insufficient heat or moving the part before the solder has set is the cause.

Miraculously, the radio worked. It was eerie hearing Radio Havana Cuba or Radio Moscow during the cold war. I kept expecting the FBI to break down my door. I got my first lesson in political "spin" hearing Cold War propaganda and thought it was funny that the tune used to identify Radio Havana was the same one the Gillette Razor Company used in their commercials. I guess since Castro had a beard, he didn't know about Gillette razors.

My buds and I raided junk parts from behind the TV repair shop. We'd strip the chassis to use for our projects. We'll talk about those parts and what they do later.

Radio wasn't our only mischief. There were home-made black powder and the "arc." The phone company used 45-volt batteries as backup power and threw them away long before they died. Each one was composed of thirty individual AA size 1.5-volt batteries, hooked in series (end to end) and encased in a tar brick. We would grab them out of the trash bin for our nefarious purposes. These were **non-rechargeable carbon-zinc batteries.**

HOW I GOT STARTED

Types of rechargeable batteries are:
Nickel-metal hydride
Lithium-ion
Lead-acid gel-cell
All the choices are correct.
Hint: Lots of rechargeable battery types so all are correct. Cheat: You don't have to recite the correct answers just know they are all correct.

If you charge or discharge lead-acid batteries too quickly, they can overheat, give off flammable gas, or explode. The gas is hydrogen. Remember the Hindenburg? An exploding car battery will also spray sulfuric acid.

Voltage is electromotive force. The basic unit is the volt. **We could measure it with a voltmeter hooked up in parallel with (across) the circuit –** from one end of the battery to the other.

You do have to be careful when measuring high voltages. **Ensure that the meter and leads are rated at the voltages to be measured.**

We knew that two of the 45-volt phone company batteries in series (end to end) are 90 volts, wouldn't it be fun to hook twenty 45 volt of them in series for 900 volts? Why stop there? We have lots. I am not sure where we stopped, but we had too many batteries for our own good.
Current flowing through a body may cause a
health hazard by:
Heating tissue
Disrupting electrical functions
Causing involuntary muscle contractions
All of the choices are correct

My buddy, Billy, struck the two wires together and pulled them apart to draw a flaming, spewing, spitting and smoking foot-long electrical arc. **The flow of**

electrons is called current. Electrical current is measured in amperes.

We could have **measured the amperes (amps) with an ammeter. An ammeter is connected in series (in line) with the circuit.** You measure flow through the circuit.

Copper is a good conductor, but there is some resistance to the flow of current even in copper wire. **A good insulator is glass.** It has a very high resistance and won't pass current.

Resistance is measured in ohms. The instrument to measure resistance is an ohmmeter.

The resistance of the wire caused the wires to consume some of the power, and the wires got so hot Billy had to drop them. The batteries heated up as well, and the entire crate of batteries melted in a pool of steaming tar. We should have used a fused line. **The component that is used to protect other circuit components from current overloads is a fuse. Don't put a 20 amp fuse in place of a 5 amp fuse or excessive current could cause a fire.**

The pool of hot tar was nothing compared to the spark that flew into a large jar of our homemade black powder. I could see my short life passing before my eyes. All the ingredients to make gunpowder were in our local hardware store, but we had never really perfected the recipe to make it explosive. I remember the hardware-store clerk glaring at us with that "I know what you boys are up to" look. I also think he knew I had listened to Radio Moscow. Russians got blamed for everything, even back then.

It was fortunate the black powder did not explode, but now we had a Fourth-of-July fountain erupting in Billy's basement. The story has a happy ending: we

HOW I GOT STARTED

didn't electrocute ourselves, the house did not burn down and, to my knowledge, Billy's parents never found out. Although, now that I think about it, I wasn't invited over for a while. That was probably seventh or eighth grade.

We were experimenters and builders – kids with curiosity. Today, we would be called "makers." What about ham radio? What was the lure? Simple. It was the first time all the stuff they were trying to teach me in school had a use. How many times have you heard adolescents complain, "I don't need to know algebra." "I'll never use chemistry." "Why do I have to learn about the ionosphere?" "Who cares where Pitcairn Island is?" And, on and on.

If you want to know how long to make an antenna, you use a little algebra. If you want to know the best time to talk to Australia, you need to understand the ionosphere. Battery chemistry will help you decide the best power source for your equipment. If you know geography, you know which way to turn your directional antenna. If you know languages, you can say "hi" to a foreigner in his own. If you know astronomy, you can appreciate meteor scatter and sun spots or know when to bounce a signal off the moon or satellite. If you learn physics, you can design an antenna. If you know history, you can appreciate Pitcairn Island. I could teach most of an entire high school curriculum based on amateur radio alone.

And what does Pitcairn Island have to do with it? Plenty. During the summer, Mom would take me to the library. I enjoyed the tall-ship swash-buckling tales with pirates and mutinies. My favorite was The Mutiny on the Bounty – a true story from 1789 with real villains and heroes. It is up to you to decide who was which.

The mutineers, led by Fletcher Christian, fled to Pitcairn Island in a remote part of the Pacific Ocean. One night, tuning the radio, I heard VR6TC on Pitcairn Island. The operator, Tom Christian was Fletcher's descendant. I "worked" him. That's ham-speak for "we made contact." I was shaking.

Tom was famous because of the Bounty story and the rarity of his location. I never heard Tom on the air again, and he became a Silent Key[3] a few years ago. Pitcairn was so remote and isolated; there were no other means to communicate with the outside world. That is why Tom and his wife had their ham licenses.

Monk Apollo in the Holy Monastery of Docheiariou at Mt. Athos (Greece) has his license for the same reason. The monastery was founded in the 10th century and is closed to the world. When one of the monks fell ill, there was no way to summon help. Monk Apollo obtained an amateur radio license to provide emergency communication. Talking to Monk Apollo is another thrill. See what you can learn from ham radio?

I wrote a "What I did last summer" essay about contacting Tom Christian for English class. My teacher didn't believe me. She said the assignment wasn't supposed to be fiction.

Months later, I received a QSL[4] card from Tom. I enjoyed rubbing her nose in it. Maybe that's why she gave me a C in English that quarter. Gloating is never a wise move.

---

[3] A deceased ham is referred to as a Silent Key, a throwback to the days of telegraph when the deceased's Morse code key went silent. You will see the call sign listed VR6TC (SK).

[4] Hams use Q signals as abbreviations. A QSL card is a postcard hams send each other as a written confirmation of their contact.

Somehow, over the years, Tom's QSL card, VR6TC, was lost, but you can see it on the wall to the top left of the map in my picture. The picture is from 1967.

The black box in the picture of my station is the receiver and the gray box to the right is the transmitter I built from a kit. Today, most hams use **transceivers: a unit combining the functions of transmitter and receiver.**

Here is Monk Apollo's card from Mount Athos. It is one of the rarer ones in my collection.

K4IA, VR6TC, SY2A – what does it all mean? The country's communications agency issues call signs. In the United States, the **FCC, Federal Communications Commission regulates and enforces the rules.** Worldwide, regulation is through the **International Telecommunications Union (ITU) which is a United Nations agency for information and communication technology issues.** Each country is assigned prefixes so you can tell a person's location from the prefix.

A valid US Technician class amateur radio call sign would be K1XXX
Hint: Technicians can request a 1x3 call sign through the vanity licensing program. First you get your assigned call, then you request a change. Remember, 3 X.

You use the call sign to identify yourself, in English, at least every 10 minutes and at the end of your conversation.

HOW I GOT STARTED

My favorite mode was and still is, CW, Morse code. There were several versions of Morse Code. **The code we use when sending CW is called International Morse code.** I liked it because no one could tell I was a kid. My squeaky voice was a dead giveaway on voice modes.

My call sign at the time was WA4TUF. When I was re-licensed in 1999, the FCC gave me KG4CVN – a real mouthful and laborious to send on CW. C and V sound alike. **FCC rules encourage the use of a phonetic alphabet for station identification. To ensure voice messages containing unusual words are received correctly, spell the words using a standard phonetic alphabet.**

Fortunately, the FCC has a vanity license program. Once you get your first FCC assigned call sign, **any licensed amateur can apply for a different call sign.**

One person can only hold one primary station license grant. Not sure why you would want more. **Your proof of possession is that your license appears in the FCC ULS consolidated database.** Hint: You're in the database. Nothing else is required.

Individuals get a call sign, but it is also possible to form a club, and the club receive a call sign. **The club must have at least four members.** Cheat: remember four to form.

INTRODUCTION TO AMATEUR RADIO

We're all familiar with radio, and it is so common today that we take it for granted. We have AM/FM and satellite radios in our car. We use cell phones and GPS navigation systems. We've come to depend on wireless Internet, wireless doorbells and wireless computer accessories. But, when Guglielmo Marconi transmitted a signal across the Atlantic Ocean in 1901, it was a remarkable feat that made world headlines. Marconi was an experimenter, and amateur radio is a continuation of that experimenter spirit.

When Titanic sank in 1912, the world awoke to the value of radio. Stations in Europe and the United States heard Titanic's powerful distress calls, but the radio operator on the closest ship had gone to bed and turned off his equipment. There was no standard procedure for handling distress calls at the time.

The answer on today's tests is that **the purpose of the Amateur Radio Service is advancing skills in the technical and communication phases of the radio art.** *Hint: Radio art? That odd phrase should stick with you as the right answer.*

Nobody knows where the term 'ham radio" came from, but you'll often hear it used to describe the Amateur Radio Service. I suspect people use "ham" because they have a hard time spelling amateur.

Governments have long recognized the value of a cadre of experienced radio operators. There are over 800,000 amateurs in the United States alone. That is a tremendous number of knowledgeable radio operators ready to bring their equipment and expertise to public service or emergency uses. All this

INTRODUCTION TO AMATEUR RADIO

expertise and equipment didn't cost the U.S. government a dime.

You can appreciate the need to regulate the radio spectrum. Without some form of regulation, your radio would sound like a raucous cocktail party and communication would be very difficult. Citizen's Band radio is unregulated and you can hear the result.

Worldwide, regulation is through the International Telecommunications Union (ITU) which is a United Nations agency for information and communication technology issues. *Hint: Don't memorize this – recognize ITU=worldwide= United Nations.*

An FCC licensed amateur station can operate in international waters on a vessel documented or registered in the United States.

You are allowed to operate in a foreign country when the foreign country authorizes it.
Fortunately, we have reciprocal treaties with many nations that automatically allow operation.

Amateurs are licensed to operate on frequencies that can travel around the world. In contrast, Citizens Band is limited to a power of 5 watts (about the same as a night light) while licensed amateurs can operate with up to 1,500 watts. That is a big responsibility, and you need to prove you can handle it.

There are three classes of amateur radio licenses. The entry level is called the Technician Class. The Technician Class license authorizes you to transmit mainly on the: **VHF (Very High Frequency 30 – 300 MHz)** and **UHF (Ultra High Frequency 300 – 3000MHz)** bands which are suitable for local communication.

(HF, High Frequency: 3 – 30 MHz) bands are sometimes called "short-wave" and are suitable for worldwide communication." Techs have some limited privileges on HF. We'll deal with frequencies and bands later but let's keep it simple for now. The dividing lines are:
3-30, HF (High Frequency)
30-300 VHF (Very High Frequency)
300-3000 UHF (Ultra High Frequency)

The General Class license authorizes you to transmit in portions of all the amateur bands, reserving a small bit of additional spectrum for Extras. The Extra Class license gives full amateur frequency privileges.

Here's a chart of Technician HF Privileges. Don't memorize the chart. It is just to illustrate a Technician is not limited to VHF and UHF:

Band	Frequencies (In MHz)	Mode
80 meters	3.525 – 3.600	CW[5]
40 meters	7.025 – 7.125	CW
15 meters	21.025 – 21.200	CW
10 meters	28.000 – 28.300	CW, RTTY[6]/data, 200 watts maximum power
10 meters	28.300 – 28.500	CW, phone, 200 watts maximum power
Above 50 MHz	All amateur privileges	1500 watts maximum power

[5] CW is Morse Code
[6] RTTY is teletype

WHO ARE YOU?

The FCC assigns call signs to radio stations. What is that? You've heard commercial radio stations identify themselves using call signs. For instance, WMAL is a popular radio station in Washington, DC. Radio amateurs also receive a unique call sign issued by the FCC. The International Telecommunications Union (ITU) allocates call sign groups to different countries. US call signs begin with a letter W, K, N, or A signifying that we are in the United States.

If you heard an amateur whose call sign began with DL, you would know he is in Germany. A call sign starting with PY would be from Brazil. There are 340 recognized "countries" each with unique call sign prefixes. I am very proud that I have confirmed contacts with 336 of the 340. We refer to foreign countries as "DX." You can learn more about DXing in my book *"How to Chase, Work & Confirm DX – The Easy Way."*

The FCC generates your first call sign in sequential order. When Billy and I got our licenses, we took the test together and received consecutive call signs. He was WN4TUE, and I was WN4TUF.

Ten numbered districts divide the United States. The "4" in my call sign signifies that it was issued for the United States fourth call district which is the southeast. The district numbering rules have become somewhat blurred and only apply to the original issue. If you move to another area, you can still keep your call sign.

If you are operating out of your area, you can use a self-assigned indicator. Examples might be KL7CC/W3. I know he is from Alaska because his call begins with KL7. The W3 tells me he is operating from the mid-Atlantic area which is the third call district.

On voice, it would sound like "KL7CC stroke W3" or "KL7CC slant W3" or "KL7CC slash W3." Stroke, slant, slash – it's all the same, and all of these choices are correct.

When our radio club operates in support of a public service event, we will often use what are called "tactical" call signs. A station might identify as "Race Headquarters." In these situations, it is more useful for everyone to know <u>where</u> you are, rather than <u>who</u> you are.

When using tactical identifiers, your station must transmit the FCC-assigned call sign every 10 minutes during and at the end of a conversation. *Hint: "Every ten minutes." This rule never changes.*

When making on-the-air test transmissions, identify the transmitting station. You should always identify.

You're required to identify with your call sign in English every 10 minutes during and at the end of your contact. The rest of the time you can speak any language you want.

You're not allowed to use secret codes. **You can transmit encoded messages when transmitting control commands to space stations or radio control craft.** Those beeps and whistles are understood by the craft. **You can transmit without on-the-air identification when transmitting signals to control model craft.**

The FCC defines a "space station" as an amateur station more than 50 km above the earth's surface. *Hint: Remember "more than 50km."*

Note that Morse code isn't secret so it is acceptable. In fact, **call sign identification for a station**

WHO ARE YOU?

transmitting phone signals can be CW or phone emissions. You will hear repeaters identify in Morse code every ten minutes.

Hams often use phonetic alphabet because it's difficult to understand the difference between letters through static crashes and interference on the radio. C, E, D, etc. sound alike, but it is very easy to tell the difference between Charlie, Echo, and Delta. **Use of the phonetic alphabet is encouraged by the FCC when identifying your station when using phone.**
Hint: The FCC "encourages" phonetics, nothing more.

Phonetics also help when talking to someone whose main language is not English. For instance, the letter "I" is pronounced "E" in Spanish. "India" cannot be mistaken for "Echo."

There are several standard phonetic alphabets, and you will get used to them. You might hear cutesy phonetics on the air, but they don't help with understanding. Kilo 4 Tango Sierra is easy to understand. Kilo 4 Tequila Sunrise is funny but not easily recognized.

WHERE CAN YOU OPERATE?

In addition to the US and territories, y**our US amateur radio license entitles you to operate from any vessel or craft located in international waters and documented or registered in the United States.** *Remember, "International waters."*

I've worked plenty of /MM stations (maritime mobile). Some cruise lines will let you operate onboard (check first). Most of the /MMs I have worked have been freighter crew members. I have also worked stations that are /AM, meaning they are "aeronautical mobile" and operating from an airplane. Don't try this on a commercial airliner.

You are also authorized to operate your amateur station in a foreign country when the foreign country authorizes it. The United States has reciprocal operating agreements with many foreign countries so you can take your Amateur Radio station and license with you when you travel.

I took a radio to Israel in 2015. I went through security at Reagan National, JFK and with the Israeli carrier, ELAL. These are the toughest security checks in the world. I had the radio with coils of antenna wire and coax in my carry-on, and no one questioned what it was. On the way back, I put the gear in checked luggage with no issues. If you travel with your radio, take a copy of your license and a magazine ad or instruction manual showing what the equipment is – just in case.

Israel (4X) has a reciprocal agreement with the US, so I was able to operate 4X/K4IA and did not need a license from the Israeli government. I was required to have a copy of my US license with me, but no one asked to see it.

WHAT DO WE TALK ABOUT?

That's a question I get asked a lot. Typically, our conversations are very much like those you would have with any person you're meeting for the first time. We talk about common interests such as our equipment, the weather, how long we've been licensed, and other mundane **topics incidental to purposes of the Amateur Radio Service or remarks of a personal nature.** No business and it wouldn't be polite or proper to ask a foreign station when they're going to get rid of their bum dictator.

Locally, you might join a roundtable with fellow commuters. Some hams meet regularly and carry on long-distance conversations for years. They may never meet face-to-face but become part of an extended family.

Amateur radio isn't for business. Here's an example: I'm on my way over to my friend Bob's house to help him with an antenna. I call him on the radio and ask if he would like me to bring over pizza. He says, sure – whatever kind you like. Another ham, who also happens to own a pizza parlor, hears me and calls. "Hey Buck, what kind of pizza do you want. I'll have it ready for you." No good, he is using amateur radio for a business purpose. Same pizza, different results.

There is an **exception for the incidental sale or trade of equipment normally used in an Amateur Radio station and not on a regular basis.** You cannot make a business of selling Amateur Radio equipment on the air.

We're amateurs and can't charge for operating a station. **An exception exists for classroom instruction at an educational institution.** The control operator is paid for being a teacher, not a ham.

WHAT DO WE TALK ABOUT?

Transmissions that contain obscene or indecent words or language are prohibited, but there is no list of banned words. You are expected to know better. The big difference between licensed Amateur Radio operators and Citizens Band operators is that the hams are usually polite and hardly ever profane. You won't be afraid to let your family listen in on the conversations.

An amateur station is authorized to transmit music only when incidental to manned spacecraft communications. Most US astronauts are ham radio operators, and they will often get "on the air" while in space. They wake up to music sent over the ham radio link.

Hams aren't supposed to engage in "broadcasting." **Broadcasting means transmissions intended for reception by the general public** as opposed to transmissions sent to a particular station. **Broadcasting, program production or news gathering is authorized where such communications relate to the immediate safety of human life or the protection of property.**

There is another exception for beacons. **The FCC defines a beacon as an amateur station transmitting for the purpose of observing propagation or related experimental activities.** You can listen for beacons to see if the band is open from their location to yours.

Amateur radio stations may make one-way transmissions when transmitting code practice, information bulletins, or to provide emergency communications.

Willful interference to other amateur radio stations is permitted at no time.

WHO CAN YOU TALK TO?

Hams talk to other hams as the frequencies allowed to us are restricted to the Amateur Radio service.

Most countries allow ham radio operation, but **FCC-licensed amateur radio stations are prohibited from exchanging communications with any country whose administration has notified the ITU that it objects to such communications.** North Korea does not allow Amateur Radio. *Hint: Ditch the overcomplicated question. If a country objects, you can't talk to a ham there.*

There are some special rules for what is called "third-party communication." **Third-party communication is a message from one control operator on behalf of another person.** For example, you might have a friend in your shack[7], and you allow him to say hello to another station, or you pass a message over the air for delivery to a third party.

A non-licensed person is allowed to speak to a foreign station if the foreign station has a third-party agreement with the US. *Hint: This is a variation on the question above. Unless a country allows, you can't do it.*

You don't need to memorize any lists, just know they are out there and look it up if the occasion arises.

---

[7] The radio shack on a boat is the room where the radios are located. "Shack" is the term used for the place you operate your radio.

WHO IS IN CONTROL?

There is a difference between a control operator and a control point. The control operator is a person, and the control point is a place.

An amateur station is never permitted to transmit without a control operator.

Only a licensed ham can be a control operator. **The class of operator license held by the control operator determines the transmitting privileges of an amateur station.**

A Technician Class control operator may only operate with those privileges allowed to Technician Class licensees. However, if an Extra Class licensee acts as the control operator, the Technician Class licensee could transmit on any frequency or mode allowed the Extra Class license. That is because the Extra Class licensee is "in control" of the station.

A Technician Class licensee may never be the control operator of the station operating an exclusive Extra Class operator segment of the amateur bands. The Technician Class licensee can't be the control operator where only Extras have privileges.

The station licensee is the one to designate the station control operator. If you have an Extra Class operator in the shack with you, designate him the control operator and you can operate in the Extra bands.

The FCC presumes the station licensee to be the control operator of an amateur station unless documentation to the contrary is in the station records. If someone is sitting at my station and

WHO IS IN CONTROL?

using my call sign, the FCC presumes I am in control. I could prove otherwise if I have a logbook showing another licensee was actually in control at the time.

When the control operator is not the station licensee, the control operator and the station licensee are equally responsible. You might turn over the reins but you are still responsible for your station.

The amateur station control point is the location at which the control operator function is performed. There are three different kinds of control: local, remote and automatic.

Local control is when the control operator is at the control point. You are there, twiddling the knobs on the radio. If you are transmitting on a handheld radio, you are in local control.

The following are true of remote control:
The control operator must be at the control point
A control operator is required at all times
The control operator indirectly manipulates the controls
All of these choices are correct.
Hint: The first two are always required, and the third describes remote operation. Two out of three means they are all correct.

Not long ago, I spoke with an amateur using a call sign in the Azores. I recognized his voice, so I asked: "Hey Martti, how's the weather in the Azores?" Martti came back, "I don't know. It is cold and snowy here. I am sitting in my living room back home in Finland." Martti was operating his remote station in the Azores over the Internet. That's a pretty amazing bit of technology considering the operator was thousands of miles away from the equipment. The control point was in Finland. My friend in Finland, **operating his**

station over the internet - an example of remote control. He was twiddling the knobs from afar.

Automatic control is when there are no adjustments made to equipment, no spinning of dials or twisting of knobs to change frequency or adjust volume. **Repeater operation is an example of automatic control**. The equipment operates automatically, and you don't have to adjust anything or be in front of it. **A station that simultaneously retransmits the signal of another on a different channel or frequency is a repeater.** *Hint: Ditch the over-complicated question and remember a repeater repeats.*

What kind of stations can automatically retransmit? Repeater, auxiliary, and space stations. They are all types of repeaters and repeaters automatically retransmit.

We will have lots more to say about repeaters later. Just know for now that **if someone uses a repeater and violates FCC rules, the control operator of the originating station is accountable.**

If your brother-in-law is cussing a blue streak in the background, and you inadvertently broadcast it over the repeater, it is your fault. You were the control operator of the originating station. The repeater is on automatic control. There is no one at the repeater to tell your brother-in-law to pipe down. You are the one in control.

BANDS AND FREQUENCIES

The abbreviation, RF, refers to radio frequency.

A radio wave is made up of electromagnetic energy. It is called an electromagnetic wave. **A radio wave has two components: an electric and a magnetic field.** You can't see it, so to me, it is just magic. A signal leaves my wire and is picked up on the other side of the world. To quote Samuel Morse's first message, "What hath God wrought?"

A radio wave travels through free space at the speed of light. The approximate velocity is 300,000,000 meters per second.

Radio waves act like alternating current. That is, the direction of flow reverses in cycles, unlike a battery. **Direct current is the name for current that flows only in one direction. Alternating current is the name for current that reverses direction on a regular basis.**

Frequency is the term describing the number of times per second that an alternating current makes a complete cycle. We measure frequency in cycles per second. The term "hertz" is a shortcut for "cycles per second." **The unit of frequency is hertz.**

The power coming out of an outlet in your house reverses 60 times per second and thus is 60 hertz – abbreviated 60 Hz. WMAL, our local AM radio station, broadcasts on a frequency of 630,000 Hz or 630 kilohertz, abbreviated 630 kHz and standing for 630 thousand hertz. A typical Amateur Radio repeater might operate on a frequency of 147,015,000 Hz or 147.015 MHz (megahertz), abbreviated MHz, meaning 147.015 million hertz.

The proper abbreviation for megahertz is MHz.
Remember Capital M and capital H; Mr. Hertz.

The electromagnetic spectrum is commonly broken down into three groups:
3 to 30 MHz HF (High Frequency)
30 to 300 MHz VHF (Very High Frequency)
300 to 3000 MHz UHF (Ultra High Frequency)

If we know the wave is traveling at 300 million meters per second, and the frequency is 144 million cycles per second, we can calculate how far the wave will travel in one cycle. **Wavelength is the name for the distance a radio wave travels during one complete cycle.**

The higher the frequency, the more often the wave reverses direction and the less distance it can travel in a cycle. Therefore, **the wavelength gets shorter as the frequency increases.**

The formula for converting frequency to wavelength in meters is "wavelength in meters equals 300 divided by frequency in megahertz."
For example, if the frequency is 150 MHz, the wavelength is 300 divided by 150 or 2 meters. The same formula works in reverse: Divide 300 by the wavelength in meters to come up with frequency. 300/2 meters = 150 Mhz. *Cheat: The only correct answer has "300 divided by."*

Let's jump now to the concept of a "band." A band is a group of frequencies. You're no doubt familiar with the AM radio band that stretches from 550 kHz to 1600 kHz. The 2 meter amateur band goes from 144 – 148 MHz.

The property of radio waves often used to identify the different frequency bands is the approximate wavelength.

BANDS AND FREQUENCIES

For example, **146.52MHz is in the 2 meter band.**
(300/146.52 = about 2) The math doesn't work out
exactly because 300 divided by 146.52 is not exactly 2
but it is close and bands refer to a range of
frequencies, so we round off to two meters.
Cheat: The other possible answers are wildly off.

An exception to the "closest answer" is for 6 meters.
**The frequency 52.525 MHz is within the 6 meter
amateur band.** 300/6 = 50 and that is the bottom of
the 6 meter band. *Cheat: 52.525 MHz is a repeating
number and stands out from the other answers.*

After a while, you'll catch on, and when someone
refers to the 2 meter band, you'll know they mean 144
to 148 MHz without thinking.

Your amateur radio license authorizes you to transmit
on certain bands. Not all frequencies are available to
amateurs, and not all amateurs can transmit on all
amateur frequencies. The higher the class license, the
more operating privileges. Most hams have a chart in
the shack because it is easier than trying to memorize
the restrictions.

MODES

The mode is the transmission type. You know about AM and FM modes from your car radio. They are different means of modulating and receiving radio signals. **Modulation describes combining speech with an RF carrier.** The RF carrier is the base frequency of the transmission. AM is amplitude modulation, and FM is frequency modulation. Amateur radio operators use many different modes.

The following are digital communications modes:
IEEE 802.11
Packet
JT65
All these choices are correct
Hint: Recognize, don't memorize. If you recognize two out of the three, the answer must be all of the above. There is more on digital modes in the chapter about computers.

NTSC is fast-scan color TV. *Cheat: C for color.* "Bandwidth" describes how much spectrum is required by a mode. The widest mode is amateur television because transmitting video requires a great amount of information. **The typical bandwidth of analog fast-scan TV transmissions is about 6 MHz.** That is the equivalent of 40,000 CW signals. *Cheat: Remember TV6.*

FM MODE

FM mode (frequency modulation), communicates information by jiggling the frequency of the transmission resulting in a signal which is relatively wide but clear and high fidelity. The frequency jiggling is called deviation.

MODES

Increased deviation makes the signal occupy more bandwidth. The amount of deviation is determined by the amplitude (loudness).

If you are told you are over-deviating, talk further away from the microphone. Your signal is too wide and might interfere with others.

FM is most commonly used for VHF and UHF voice repeaters.

The approximate bandwidth of a VHF repeater FM phone signal is between 10 and 15 kHz.

Because the signal is wide, it can carry a lot of information. **Frequency modulation (FM) is most commonly used for VHF packet radio transmission.**[8]

AM MODE

In the AM Mode (amplitude modulation), information is communicated by jiggling the amplitude of the radio wave. The audio information varies the power of the radio frequency wave, and the result is the center frequency called the carrier and a sideband on either side of the carrier varying in amplitude in accord with the audio. AM was the phone mode when I was getting started. But, AM requires lots of power and heavy-duty equipment to handle that power.

SINGLE SIDEBAND MODE (SSB)

Single sideband sounds like Donald Duck until tuned in properly. But, SSB is a superior mode because it requires less power and takes up less bandwidth. Eventually, SSB won out over AM and today you will

---

[8] Packet radio is data transmission. It got that name because the data is transmitted in clumps or packets.

only hear occasional AM from boat-anchor enthusiasts[9] – those folks with nostalgia for vintage radios, vacuum tubes and the good-old days.

Single sideband (SSB) is a form of amplitude modulation (AM). Look at the diagram. The bottom waveform is an AM signal. To make SSB, 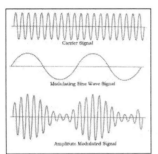 filters in the transmitter strip away the carrier and one of the sidebands leaving only a single sideband. The remaining sideband could be either on the upper or lower side of the carrier. You'll hear this referred to as upper sideband or lower sideband. By convention, upper sideband is used on higher frequencies.

The voice mode used for long distance weak signal contacts on VHF and UHF is SSB. Single sideband carries better than the other voice modes

Upper sideband (USB) is normally used for 10 meter HF, VHF, and UHF single sideband communications. *Cheat: The answer to remember for the test is "upper." There is no test question with a correct answer of "lower."*

The approximate bandwidth of a single sideband (SSB) voice signal is 3 kHz. (3000 Hertz) SSB is quite a bit narrower than FM (10-15kHz) or AM (6kHz). *Cheat: Remember SSB has three letters.*

The advantage of single sideband over FM for voice transmissions is SSB signals have a narrower bandwidth. More signals can fit in a given part of the radio spectrum. Narrower signals also

---

[9] A "boat-anchor" is a heavy old piece of equipment suitable for anchoring a boat.

mean you can use narrow filters to reduce noise and interference.

Our early radios had very poor selectivity. That meant you heard many signals, and it was difficult to pick out the one you wanted to hear. **The ability to discriminate between multiple signals is called selectivity.** To get more selectivity, we use filters to reduce the bandwidth of the receiver.

You would use a filter 2400 Hz wide for minimizing noise and interference on SSB. That is 2.4 kHz and just slightly narrower than the signal.

SSB voice modulation is most often used for long distance or weak signal contacts on the VHF and UHF bands. It is the primary audio mode on the HF bands as well.

If the voice pitch seems too high or too low, you would use the receiver RIT (Receiver Incremental Tuning) or Clarifier control to fine tune. The voice pitch is off because you are slightly off frequency. The RIT or Clarifier fine-tunes the receive frequency without changing your transmit frequency.

CW MODE (MORSE CODE)

CW, which stands for continuous wave or carrier wave, is the most basic of digital modes. The carrier is turned on and off making the dots and dashes we read as Morse code. There were several versions of Morse code, but **International Morse code is used when sending CW in the amateur bands.**

You are not allowed to use codes or ciphers that hide the meaning of a message except when transmitting control commands to space stations or radio remote control craft. Control commands are not ciphers to hide the meaning. Morse code is widely used and easily decoded so it is not a cipher.

In comparison to SSB and FM, **CW emission has the narrowest bandwidth. The approximate maximum bandwidth required to transmit a CW signal is 150 Hz.** That means almost 20 CW signals can fit in the same space as one SSB signal.

The typical filter used for CW is 500 Hz wide. The narrower filter cuts down on interference and noise outside the passband.

An electronic keyer assists in the sending of Morse code. Instead of pumping a key up-and-down, you push a lever one way to generate a string of dots and another way for dashes. Much faster.

The advantage of having multiple bandwidth choices is it permits noise or interference reduction by selecting a bandwidth matching the mode. You select narrower filters for narrower modes.

BANDPLANS

The FCC and ITU dictate the bands of frequencies available for amateur operation. The rules restrict Technician Class licensees to certain frequency bands and parts of bands for different modes.

As a Technician class operator, you have HF phone privileges on 10 meters only. A Technician class operator has HF, RTTY, and data privileges on 10 meters only. *Cheat: The answer to both questions about HF is 10 meters.*

In the HF bands, a Technician is limited to 200 watts output power. Above 30 MHz, a Technician can use up to 1500 watts.

The VHF frequencies limited to CW only are 50 MHz to 50.1 MHz and 144.00 MHZ to 144.1MHZ. *Cheat: This is the only answer with VHF frequencies.*

Between 219 and 220 MHz, is limited to fixed digital messaging forwarding systems only. *Hint: Just remember "fixed digital."*

For the most part, amateur bands are only available to amateur radio operators. However, we share some bands with other services. **When the Amateur Radio Service is secondary, you may find non-amateur stations and must avoid interfering.** *Hint: We are secondary, and you must defer to the primary user and not interfere.*

You should not set your transmit frequency to be exactly at the edge of an amateur band: To allow for a calibration error in the transmitter display frequency. The display may not be accurate. **So that modulation sidebands do not extend beyond the band edge.** Your transmission is wider than you think

To allow for transmitter frequency drift. Your radio might drift as it heats up.
All these choices are correct.
Hint: There are lots of reasons not to get too close to the edge. All the choices are correct.

SSB phone may be used in at least some portion of amateur bands above 50 MHz. *Hint: Something is better than nothing.*

There are also voluntary guidelines beyond the privileges established by the FCC. These are voluntary "band plans." An example of a voluntary band plan is a national calling frequency.

The national calling frequency on the 2 meter band is 146.520 MHz. That is the watering hole where everyone goes to look for a contact. *Cheat: When the test asks for a "national calling frequency" you only need to recognize one answer: 146.520 MHz.*

There are national calling frequencies on other bands but you don't need to remember them for the test.

PROPAGATION

Propagation is the term used to describe the distribution of radio waves. Radio waves travel in straight lines, and unless they bounce off something, communication is limited to what is called "line of sight." The curvature of the Earth blocks radio waves. Buildings, mountains or other physical obstacles can also block radio waves.

When signals bounce off multiple buildings or other objects, it can cause distortion. Just like it is hard to understand music or conversation in a big boomy room with lots of echoes, radio echoes can be distracting. These echoes are "multi-path" distortion.

If the other station reports your signal was strong a moment ago but is now weak or distorted, try moving a few feet or change the direction of your antenna as reflections may be causing multi-path distortion. You may have moved into the shadow of a building or passing truck.

The term, picket fencing, is commonly used to describe the rapid fluttering sound sometimes heard from mobile stations that are moving when transmitting. Picket fencing is a rapid in-and-out flutter usually caused by multi-path distortion.

The part of the atmosphere called the ionosphere enables the propagation of radio signals around the world. The ionosphere is clouds of ions energized by the sun. Signals bounce (refract) off the ionosphere and are reflected back to Earth; a phenomenon called "skip." In some cases, the signals will bounce off the ionosphere back to Earth back to the ionosphere and back to the Earth again for what is called "multi-hop" propagation. Signals can travel around the world in this manner.

The peak of the sunspot cycle makes long-distance communication possible on six or ten meters. The peak of the sunspot cycle brings more solar energy to the ionosphere and better propagation. *Hint: The other answers are all UHF bands which do not refract off the ionosphere.*

The advantage of HF vs VHF and higher frequencies is long distance ionospheric propagation is more common on HF. The higher frequencies don't bounce off the ionosphere.

The 10-meter band is around 28 MHz, and you can operate phone, CW and data modes there with your Technician license. Please don't think your Technician license limits you to shack-on-a-belt HTs and repeaters. As a Tech, you have some privileges on HF bands and can experience worldwide communications with a modest station.

Direct UHF signals (not on a repeater) are rarely heard from stations outside your local coverage area because UHF signals are usually not reflected by the ionosphere. VHF and UHF frequencies pass through the ionosphere and out into space. That is why VHF and UHF frequencies are used to communicate with satellites.

VHF and UHF radio signals usually travel somewhat farther than the visual line of sight distance between two stations because the Earth seems less curved to radio waves than to light. *Cheat: That sounds bizarre, so it's an easy test answer to remember.* Air bends the radio wave over the horizon causing the phenomenon.

The range of VHF and UHF signals might be greater in the winter because there is less

PROPAGATION

absorption by vegetation. Trees can absorb VHF and UHF signals.

Fog and light rain have little effect on 10 and 6 meters. HF is not much affected by weather.
At microwave frequencies, precipitation can decrease range. "Microwave" is another way of saying "UHF."

Signals can also bounce off buildings or bend around a sharp corner. **Knife-edge diffraction might cause signals to be heard despite obstructions between the transmitting and receiving stations.** The term "knife-edge diffraction refers to signals that are partially refracted around sharp edges of solid objects. The term "refracted" is another way of saying "bent."

Radio waves can refract off more than just the ionosphere and buildings. One example is called "tropospheric ducting." **Tropospheric ducting is caused by temperature inversions in the atmosphere.** The radio waves are trapped between layers of different temperature air and bounce along until they squirt out some ways away.

The phenomenon responsible for allowing over the horizon VHF and UHF communications to ranges of approximately 300 miles on a regular basis is tropospheric scatter.

Another propagation example is called "auroral reflection." Signals traveling near the North or South Pole may reflect off the Aurora Borealis, or northern lights. **Signals exhibiting rapid fluctuations of strength and often sounding distorted is a characteristic of signals received via auroral reflection.** On HF, signals traveling over the North Pole from Asiatic Russia will often sound spooky, watery or fluttery. This sound is classic aurora.

When meteors pass through the atmosphere, they leave behind a short-lived trail of ionized gas. Radio waves will bounce off the ionized gas. The 6 meter band can be very active during times of known meteor showers.

The band best suited to communicating via meteor scatter is 6 meters. *Cheat: "Meteor" has 6 letters, 6 meter = meteor.*

The ionosphere is divided into several layers commonly referred to as the D, E, and F layers. These layers are energized by the sun and exhibit different characteristics depending on the time of day, time of year and the amount of energy coming from the sun.

The E layer has a component that can be unpredictable and sporadic ("Sporadic E"). **Occasional strong over-the-horizon signals on the 10, 6, and 2 meter bands are sporadic E.** *Cheat: If you see "sporadic" in an answer, it is right.*

BE SAFE, STAY SAFE

A fuse protects circuits from current overloads.
A thin metal strip in the fuse overheats and melts, breaking the connection. **The purpose of a fuse is to interrupt power in case of an overload.**

Equipment powered from a 120V AC power circuit should always include a fuse or circuit breaker on the hot conductor.

Don't put a 20-ampere fuse in place of a 5-ampere fuse or excessive current could cause a fire. You lose protection as this would allow 20 amps to flow in a circuit only designed to handle 5 amps. The overload would cause the circuit to overheat.

The following question assumes you are talking about the plug in a wall socket.
To guard against electrical shock:
Use three-wire cords
Connect to a common safety ground
Use ground fault interrupters.
All the answers are correct.
Hint: There are lots of ways to guard against shock.

The green wire on a plug goes to safety ground.
Hint: Green for ground. That is the lone pin on an electric plug.

When putting up a tower, look and stay clear of overhead wires. Overhead power wires carry lethal voltages. If you contact an overhead wire with a metal antenna support, you will be seriously injured, if not killed. Look up before you put up.

To determine the minimum safe distance from a power line, make sure if the antenna falls, it can come no closer than a minimum of 10 feet. If your pole is 30 feet long, be sure you are at least 40 feet from the power line. Measure and put something on the ground, so you don't wander into danger.

Don't attach an antenna to a utility pole because the antenna could contact the high-voltage power lines.

Use a carefully inspected climbing harness and safety glasses when climbing a tower. The climbing harness will keep you from falling because it attaches you to the tower.

Wear a hard hat and safety glasses at all times when any work is being done on the tower. A tool falling from a tower can cause serious injury. **It is never safe to climb a tower without a helper or observer.** Use the buddy system.

Never climb a crank-up tower unless it is retracted or mechanical locking devices have been installed. Riding a ladder down looks funny in National Lampoon's "Christmas Vacation," but it is very dangerous.

A gin pole is an extension that is used to lift tower sections or antennas above the top of the tower. It is a temporary extension above the tower.

A safety wire through a turnbuckle is to prevent loosening of the guy line. The wire keeps the turnbuckle from turning due to wind vibration.

Local electrical codes set grounding requirements. Grounding and house wiring are local electrical code issues. The FCC is not involved.

The lowest impedance (resistance) to RF signals is flat strap. RF travels on the surface, a phenomenon known as "skin effect." Flat strap has more surface area than round wire, so it is more effective for conducting and grounding RF.

Proper grounding for a tower is a separate eight-foot rod for each leg bonded to the tower and each other. *Hint: The more the merrier.*

When installing ground wires on a tower for lightning protection, ensure that all connections are short and direct. For lightning protection, sharp bends must be avoided. Short, straight and direct. Bends impede lightning's direct path to ground. **Mount all devices for lightning protection to a metal plate that is in turn connected to an external ground rod.**

You should bond all external ground rods or earth connections with heavy wire or conductive strap. There should be no opportunity for different voltages between different parts of your station. Even a slight difference in resistance to ground can cause a lot of current to flow through your equipment with damaging results.

VHF and UHF radio are non-ionizing radiation. RF radiation differs from ionizing radiation (radioactivity) because RF does not have sufficient energy to cause genetic damage. No evidence yet that RF causes cancer.

But RF exposure can be dangerous. We heat food with RF in a microwave oven, and our bodies can be heated by RF as well.
The factors affecting RF exposure are:
Frequency and power level
Distance from the antenna
Radiation pattern of the antenna.
All these choices are correct.

Exposure limits vary with frequency because the human body absorbs more RF at some frequencies than others.

50 MHz has the lowest value for Maximum Permissible Exposure Limit. Meaning, 50 MHz, the 6 meter band heats our body the most.

If you run more than 50 watts at the antenna on VHF frequencies, you are supposed to do an RF exposure evaluation. *Cheat: The answer, both here and above, is "50." 50 watts and 50 MHZ.* **To stay in compliance, you should re-evaluate whenever you change equipment.**

An acceptable way to check exposure is using: FCC Bulletin Computer modeling Measurement of field strength using calibrated equipment All these choices are correct. *Hint: Lots of ways to check, so all the choices are correct.*

To prevent exposure, try relocating the antennas. Moving further away from the antenna will reduce your exposure.

Duty cycle is the percentage of time the transmitter is transmitting. Duty cycle is a factor because it affects the average exposure.

If the limit is 6 minutes and you are on 3 and off 3, you can be exposed safely for two times as long. That would be a 50% duty cycle, and your body has time to cool during the off times. Therefore, you could be exposed safely for twice as long.

A person touching your antenna might get a painful RF burn.

RADIO OPERATION

Now that we've covered basic concepts of frequency, mode, and propagation, let's start operating.

A ham shack from the 60's would have included a transmitter and a receiver – two separate boxes. You can see that in my shack picture from 1967 on page 16. Today's radios are **transceivers which combine a transmitter and receiver** in one box. The two automatically track each other with one knob.

If you are transmitting and receiving on the same frequency, it is simplex communication. *Hint: One frequency is simple(x).*

Your microphone has a **PTT (Push To Talk) switch that switches between receive and transmit.**

Modern radios will let you **store the frequency in memory channels for quick access.** This is much easier than spinning the dial to get to your favorite frequencies. **You can also enter your frequency by typing it on a keypad or use the VFO knob.** VFO stands for "variable frequency oscillator," and you would be turning the knob to change frequency.

The scanning function of an FM transceiver is used to scan through a range of frequencies to check for activity. *Hint: A scanner scans.*

An oscillator is a circuit that generates a signal at a specific frequency.

Your HT, hand-held or handy-talkie is a transceiver. The short rubberized antenna that came on top of your HT is called a "rubber duck." **A disadvantage of the "rubber duck" antenna supplied with most handheld radio transceivers is it does not**

transmit or receive as effectively as a full-sized antenna. It is handy but too short.

A good reason not to use a handheld transceiver inside your car is that signals may not propagate as well due to the shielding effect of the vehicle. The metal car body acts like a shielded cage trapping the signals inside. Mount an antenna outside the vehicle.

The device that increases the low power output from a handheld is an RF power amplifier. *Hint: Increase power with an amplifier.*

The function of the SSB/CW-FM switch on an amplifier is to set the amplifier for proper operation in the selected mode. It doesn't change the mode because amplifiers can't do that, they just amplify. *Hint: "Set for proper operation."*

Crackling static or white noise can be very fatiguing and is guaranteed to drive spouses crazy. **The squelch control mutes the receiver output noise when no signal is present.** *Hint: Tie the words "squelch" and "mute" together.*

A sub-audible tone transmitted along with normal voice audio to open the squelch of a receiver is called CTCSS. Coninuous Tone-Coded Squelch System. The tone activates the receiver eliminating unwanted conversations.

There are a dozen or so "Q" signals in common use. You only need to know these two for the Technician test. Q signals are both shorthand and a universal language. If I tell a Russian "QRM,", I am telling him **"I am receiving interference" QRM.** He doesn't speak English, and I don't speak Russian, but the point is understood.

RADIO OPERATION

Another example is **I am changing frequency: QSY.**

Contacting as many stations as possible during a specific time is contesting. In a radio contest, you send only the minimum information needed for proper identification and the contest exchange. The idea is to contact as many stations as possible not to chew the fat. The contest exchange is the information you send and receive to validate the contact. The contest rules set the exchange, which might be a serial number and your state.

There are several contests every weekend, and some hams are fierce competitors. It is possible to work all 50 states or 100 countries in one weekend, qualifying you for a Worked All States (WAS) or DXCC (DX Century Club) award from ARRL, the American Radio Relay League. ARRL calls itself the national association for amateur radio (arrl.org).

"DX" is shorthand for distance. Since my contact with Tom Christian, I have been an avid DXer, chasing new countries. It got so bad one summer, Mom kicked me out of the basement and told me to play outside. She said my skin was turning blue-green from lack of sunlight.[10]

A Grid Locater is a letter-number assigned to a geographic location. The world is divided into "Maidenhead Grid Squares." Someone might not know the location of Fredericksburg, VA but they can find Grid Square FM18. VHF and UHF contests often use Grid Locator squares as part of the contest exchange.

---

[10] Check out my book "How to Chase, Work & Confirm DX – the Easy Way." It is available from Amazon.

HOW TO START A CONVERSATION

If you are not using a repeater, you would call CQ.
CQ means calling any station.

On a repeater, in place of CQ, just say your call sign. "K4IA listening."

To call another station on the repeater, if you know his call sign, say the other station's call sign, then identify with your call sign. "W3ABC this is K4IA."

If two stations are on the same frequency and interfere, use common courtesy, but no one has an absolute right to a frequency. It is easier to spin the dial than to get in an argument. You could tell your partner: "I'm going to QSY (change frequency) up 2 kHz."

To respond to a CQ, transmit the other station's call sign followed by your call sign. W3ABC calls "CQ, CQ, CQ. This is W3ABC, Whiskey 3 Alpha Bravo Charlie." I respond, "W3ABC this is K4IA, Kilo 4 India Alpha." *Hint: His call sign first because you want to get his attention.*

Before choosing an operating frequency for calling CQ:
Listen first to be sure no one else is using the frequency
Ask if the frequency is in use
Make sure you are in your assigned band
All of these choices are correct.

EXTENDING YOUR RANGE WITH REPEATERS

You can get on the air with an HT for less than $50 and be able to talk line-of-sight, a couple of miles on simplex, and much further through the use of a repeater. **Simplex communication is the term used to describe an amateur station transmitting and receiving on the same frequency. Simplex channels are designated in the VHF/UHF band plans so stations can communicate directly without using a repeater.** No need to tie up a repeater when you can speak directly.

We already discussed the fact that 2 meter (VHF) radio waves do not bounce off the ionosphere. They will bounce off buildings, and if you're high enough or conditions are just right for tropospheric ducting, you might be able to communicate over a few hundred miles. But we need some help for reliable local communication. That's where a repeater comes in handy.

A repeater simultaneously retransmits a signal on a different channel or channels. The repeater "repeats" the signal to relay over a longer distance.

Since 2 meter communication is line-of-sight, the higher you are, the further you can see and the further the radio waves will reach. A repeater antenna will be on a very high site– usually atop a tall building, water tower or mountain.

In addition to the tall antenna, the transmitter portion of the repeater transmits with more power than the typical handheld or a radio mounted in your car. The combination of extra height and power stretch the range out to perhaps 10 miles for an HT or 50 miles for a full-power mobile radio. Those are the basics of a repeater.

Repeater, auxiliary, and space stations can automatically retransmit. They are all repeaters and repeaters automatically retransmit. An auxiliary station is a repeater for a repeater.

A linked repeater network is a network of repeaters where signals received by one are repeated by the others. This expands coverage greatly.

Usually, a local club builds and maintains the repeater with member dues and that is a good reason to support your local club.

To prevent interference, a Volunteer Frequency Coordinator recommends receive/transmit channels and other parameters. *Hint: A coordinator "coordinates."* **A Frequency Coordinator is selected by local amateur operators whose stations are eligible to be repeater or auxiliary stations.** *Hint: Ditch the complications and remember he is selected by amateurs.*

Simplex communication is the term used to describe an amateur station transmitting and receiving on the same frequency. A repeater operates in "duplex" mode. Since a repeater is receiving and transmitting at the same time, it makes sense that it cannot do both on the same frequency. It would interfere with itself. The repeater listens and talks on two different frequencies simultaneously to prevent self-interference.

"Repeater offset" is the difference between a repeater's transmit and receive frequencies. A common repeater offset in the 2 meter band is plus or minus 600 kHz. For example, my local

REPEATERS

repeater receives on 147.615 MHz and retransmits on 147.015 MHz. To access the repeater, you would tune your receiver to 147.015 so you can hear the repeater, and you would set the offset on your radio to transmit +600 kHz so the repeater can hear you.

A common use of the "reverse split" function of a VHF/UHF transceiver is to listen on a repeater's input frequency. If you can hear the input, you might be able to go to simplex communication and not tie up the repeater. Or, the station might be closer to you than he is to the repeater and you can hear him better on the input frequency. **If a station is not strong enough to keep the repeater's receiver squelch open, you might be able to receive the station's signal by listening on the repeater input frequency.** *Cheat: Any answer containing "repeater input frequency" is correct.*

A common repeater frequency offset in the 70 cm band is plus or minus 5 MHz. The 70 cm band is much wider than the 2 meter band, and there is plenty of room to spread out, so the repeater offset is greater.

A way to enable quick access to a favorite frequency on your transceiver is to store the frequency in a memory channel. Your radio will have memory channels, so once you set frequency and offset data in the memory, you do not have to re-enter it every time you turn on your radio or switch to a different repeater.

To prevent interference among repeaters, many repeaters use a sub-audible tone as a "key" for access. It is sub-audible meaning too low in pitch for you to hear. You program the sub-audible tone in a memory channel along with the other repeater data.

Continuous Tone Coded Squelch System (CTCSS) is the term used to describe the use of the sub-audible tone transmitted with normal voice audio to open the squelch of a receiver. *Remember, CTCSS opens the squelch.*

If you're trying to access a repeater and failing, chances are you may not have the proper CTCSS, DTMF, or audio tone for access. DTMF is the tone generated by your keypad. Some repeaters require a DTMF code. Don't let all those fancy names fool you. They are just another way of saying you may need a special tone to unlock the repeater.

If you can hear a repeater's output but can't access it, a reason might be:
Improper transceiver offset
The repeater may require a CTCSS tone'
The repeater may require a DCS tone
All of these choices are correct
Cheat: If you recognize two, the answer is all of the above.

APPS for iPhone and Android use the GPS in your phone to locate nearby repeaters and provide access information.   Look for an APP called "Repeater Book."

DMR (Digital Mobile Radio or Digital Migration Radio) describes a system for time-multiplexing two digital voice signals on single repeater channel. *Hint: Ditch the over-complicated answer and know two people can talk on the same channel with DMR.*

A "talk group" on a DMR digital repeater is a way for groups of users to share a channel without being heard by other users of the channel. *Hint: Don't over complicate it, when you join a group, you only hear transmissions for that group.* **To join a**

REPEATERS

repeater's talk group, program your radio with the group's ID or code. *Hint: Program in the key.*

Repeaters can also provide a gateway to the internet. **A station that is used to connect other amateur stations to the internet is called a gateway.**

The Internet Radio Linking Project (IRLP) connects amateur radio systems via the internet. The tones used to control repeaters linked by the IRLP protocol are DTMF. They are generated by your transceiver keypad just like the tones on your telephone.

You can also access some IRLP nodes by using DTMF signals. DTMF stands for Dual Tone Multi-Frequency. Dial in the "phone number" of a distant repeater, and your signal will pop out of that repeater – even on the other side of the world. Dial 5600 and your signal will travel over the internet to a repeater in London.

Echolink is another internet linking protocol. **Before you can use the Echolink system, you must register your call sign and provide proof of license.** *Hint: You need to register and prove.*

Voice Over Internet protocol is a method of delivering voice over the Internet using digital techniques. Skype is VoIP. *Hint: the name tells you the answer.*

You can find lists of active nodes that use VoIP By subscribing to an online service From on line repeater lists maintained by the local frequency coordinator From a repeater directory All of these choices are correct.

OPERATING SIMPLEX

Let's get on the air. Assume you have your radio set up and you want to talk to somebody. Listen first and select a clear frequency. You don't want to cause harmful interference to another user.

It helps if you start off on a frequency where other people commonly listen. These are called "calling frequencies." **The national calling frequency for FM simplex operations in the 2 meter band is 146.520 MHz.** *Cheat: The test only asks one calling frequency, so memorize 146.520.*

If you are looking to establish a contact on HF, you can call CQ. **The meaning of the procedural signal "CQ" is calling any station.** You would say, "CQ CQ CQ this is K4IA calling CQ and listening." We don't use CQ on a repeater.

Some folks might recommend a longer CQ sequence, and I won't argue with them, just don't overdo it. If you go on and on and on, whoever's listening will get tired of waiting and tune away. If you don't get a response the first time, call again. It's better to use several short CQ calls than one long one.

Before you call CQ, listen if the frequency is already in use.   After listening, you still can't be sure the frequency is not in use because you can't always hear both sides of a conversation. It is proper to ask, "Is this frequency occupied?" before you call CQ.

Before choosing an operating frequency for calling CQ:
Listen first to be sure no one else is using the frequency
Ask if the frequency is in use
Make sure you are in your assigned band
All of these choices are correct.

OPERATING SIMPLEX

When responding to a CQ, you should transmit the other station's call sign followed by your call sign. His call goes first to get his attention. Like this: "W3ABC this is K4IA. How copy?" If you don't know him, or if conditions are less than ideal you would use phonetics, "W3ABC this is K4IA, Kilo 4 India Alpha." You don't need to say his call in phonetics. I am sure he knows his call. It's your call he might have trouble understanding. Once he acknowledges you, you've made contact and can start your conversation.

You identify every ten minutes and at the end of your communication (conversation).

Call sign identification for a station transmitting phone signals can be CW or phone emissions. You will hear your local repeater identify in Morse code like this, "K4TS/R" K4TS is the call sign, /R indicates the station is a repeater.

EMERGENCY

Amateur radio bills itself as the communication system of last resort: "When all else fails."

A net is a group of stations operating together. "Net control" is the leader or emcee. **To get the attention of a net control station in an emergency, begin your transmission by saying "Priority" or "Emergency" followed by your call sign.** Use the same protocol on a repeater if you come across an accident and want to summon help.

The term "NCS" refers to Net Control Station.

Once you have checked into a net, remain on frequency without transmitting until asked. If you are not handling emergency traffic, stand by and wait to be called. Once you have checked in, net control knows you are there and will call you if needed. In the meantime, stay out of the way.

You can **operate outside of frequency privileges of your license class in situations involving the immediate safety of human life or protection of property**. You can use any band or mode if needed in of an emergency. "Emergency" is always an exception to the rules as in, you can "broadcast" in an emergency.

There are two groups that organize for emergency communications: RACES and ARES.

RACES is:
A radio service for emergency management and civil defense
A radio service using amateur stations for emergency management or civil defense
An emergency service using amateur operators certified by a civil defense

EMERGENCY

All the above choices are correct.
Hint: RACES does a lot of things related to civil defense so, all are correct.

ARES is also an organization for emergency communications. **ARES is licensed amateurs who have voluntarily register their qualifications and equipment for communications duty.** *Cheat: ARES has an R in it and you register.*

What RACES and ARES have in common is that both may provide communications during emergencies.

Our local club is ARES affiliated and holds a weekly net on the repeater. The purpose of the net is to familiarize operators with net operations, and give them a chance to test their equipment and skills. Check into your local ARES net. It will be good training and help you get over your initial "mic fright."

"Traffic" refers to formal messages exchanged by net stations.

Good traffic handling is passing messages exactly as received. That is word-for-word, not interpreted, condensed or editorialized. If the message is, "There are 50 people in the Stafford emergency shelter," you don't pass the message as, "There are a ton of people in Stafford."

One way to ensure a voice message containing unusual words is copied correctly is to spell the words using a standard phonetic alphabet. Using the standard phonetic alphabet is also the proper way to identify your call sign. **The FCC encourages the use of a standard phonetic alphabet** but doesn't publish one.

You don't need to memorize these, but here is the word list adopted by the International Telecommunication Union:

A--Alpha	**J**--Juliett	**S**--Sierra
B--Bravo	**K**--Kilo	**T**--Tango
C--Charlie	**L**--Lima	**U**--Uniform
D--Delta	**M**--Mike	**V**--Victor
E--Echo	**N**--November	**W**--Whiskey
F--Foxtrot	**O**--Oscar	**X**--X-ray
G--Golf	**P**--Papa	**Y**--Yankee
H--Hotel	**Q**--Quebec	**Z**--Zulu
I--India	**R**--Romeo	

Messages have a preamble. **The preamble contains the information needed to track the message as it passes through the system.** Think of it as the "to" and "from."

Another part of a message is the "Check". **Check is a count of the number of words in the text portion of the message.** If the "check" is 27 and you only count 25 words, you missed something.

SATELLITES

In 1957, the Soviet Union launched the first satellite into space. It wasn't much, just a metal ball 2 feet in diameter. Sputnik broadcast a simple Morse code message back to Earth: di di di dit di dit. "HI."

A satellite beacon is a transmission from a satellite that contains status information. Telemetry is the one-way transmission of measurements. **Telemetry information typically transmitted by satellite beacons is information about the health and status of a satellite.** The speed of the dits told Sputnik's temperature and pressure. **Anyone who can receive telemetry signal may do so.** There are no restrictions on receiving data.

The mode of transmission used for satellite beacons is
RTTY (teletype)
CW (Morse code)
Packet (short bursts of digital data)
All of these choices are correct

More than 70 Amateur Radio satellites have launched. The Amateur Radio satellites act as repeaters in space. **LEO means the satellite is in Low Earth Orbit.** That is between 100 and 1200 miles.

When a satellite acts as a repeater, your signal goes up to the satellite to be rebroadcast down. In the process, your signal can be heard over thousands of miles – the entire portion of the Earth's surface that can "see" the satellite. **Repeaters, auxiliary and space stations are all repeaters and can automatically retransmit signals.**

You can talk to the International Space Station if you have a Technician or above license. Many of the astronauts are hams, and they have ham equipment on board. Transmissions from the ISS are on 145.800 MHz, in the two-meter ham band.

Satellites move quickly. **Satellite tracking programs provide maps showing:**
The position of the satellite
Time azimuth and elevation
The Doppler shift.
All choices are correct
Hint: You need a lot of information to track a satellite, so all the choices are correct.

The inputs to a satellite tracking program are the Keplerian elements. These are equations and solutions that track the satellite as it orbits the rotating Earth.

Listening is further complicated by spin fading and Doppler shift. **Spin fading is caused by rotation of the satellite and antennas.** When the satellite spins, the antennas point away from Earth and the signal drops.

Doppler shift is an observed change in frequency caused by the relative motion between the satellite and Earth station. Just like the pitch of a siren rises as it comes toward you and falls as it moves away, the perceived frequency of the satellite's signal will change.

UHF signals are usually not reflected by the ionosphere. The signal goes right through. That is why satellite communication uses high frequencies.

A satellite operating in U/V mode is using an uplink in the 70 centimeter band and a downlink in the 2 meter band. Think of it as a repeater with an offset that is using two different bands. *Hint: 70 centimeters is UHF and VHF is 2 meters.*

SATELLITES

The control operator of a station communicating through a satellite must have license privileges that allow him to transmit on the satellite uplink frequency. *Hint" You are always required to have privileges to transmit on a given frequency.*

If you use too much power on the uplink, you can block access by other users. You overload the satellite's receiver.

A good way to judge whether your uplink power is too low or too high is your signal strength on the downlink should be about the same as the beacon. The beacon is the signal generated by the satellite to identify itself. Your downlink (received) signal should be about the same strength.

COMPUTERS

After SSB and repeaters, the next great breakthrough in Amateur Radio has been the integration of computers and radios. One use is to connect the radio with a logging program. The computer reads your frequency and mode and puts it in the log for you. The picture on the back cover is me operating Field Day.[11] You can see the notebook computer figures prominently.

A computer might be used as part of an amateur radio station
For logging contacts
For sending and receiving CW
For generating and decoding digital signals
All of these choices are correct
Hint: Lots of uses.

Digital communications modes include:
Packet,
IEEE 802.11
JT65
All of these choices are correct

Digital modes work because **your computer sound card provides audio to the microphone and converts received audio to digital form** for display on the screen.

The connections between a voice transceiver and a computer would be: receive audio, transmit audio and push-to-talk (PTT).

---

[11] Field Day is an annual exercise the last full weekend in June. Hams go to the field and operate with emergency power seeing how many contacts they can make. My personal goal is 1000 contacts in 24 hours. It is a combination of club picnic, campout and operating. Check with your local club. They are bound to have something planned and will welcome you.

COMPUTERS

The computer sound card port connected to a transceiver's headphone or speaker output would be the microphone or line input. *Hint: "Out" from the transceiver goes to "in" on the computer.*

Packet is a digital system that sends bursts of data. **Included in a packet transmission are a: Header with the call sign of the station to which the information is sent Check sum which permits error detection Automatic repeat request in case of error All the answers are correct** *Hint: Packet is self-correcting, so all the answers are correct.*

An ARQ transmission system detects errors and sends a request to retransmit. (Automatic Repeat Query). If the numbers don't add up, the receiving station automatically asks for a repeat.

APRS is another use for packet. **APRS means Automatic Packet Reporting System.**

An application of APRS is to provide real-time tactical digital communications in conjunction with a map showing the locations of stations. Data to the transmitter for automatic position reports is supplied by a GPS Global Positioning System receiver. *Hint: Ditch the complications, APRS reads your location from a GPS and broadcasts it. Others can receive the information and read your location from a computer generated map.*

Your computer sound card can also generate **PSK, which stands for Phase Shift Keying**. *Hint: The key word is "phase."* PSK is a popular digital mode on HF.

WSJT is a suite of digital protocols that support very weak signal work. **The following activities are supported by digital mode software in the WSJT suite:**
Moonbounce or Earth-Moon-Earth
Weak signal propagation beacons
Meteor scatter
All of these choices are correct
Hint: All are weak signal activities.

One program in the WSJT suite is FT8. **FT8 is a digital mode capable of operating in low signal-to-noise conditions that transmits at 15-second intervals.** Hint: Just recognize FT8 operates at "low signal levels" or "every 15 seconds."

Digital modes are very robust, and your computer can decode signals that are barely audible. As a Technician, you have data privileges on 10 meters, where you can operate worldwide using digital modes.

One last digital question asks about Broadband-Hamnet ™, also referred to as a high-speed multi-media network. Wireless computer routers operate in part of an amateur band. Some clever hams figured out how to reprogram the routers to run more power and connect to each other over broad "mesh" networks. **Broadband Hamnet is an amateur-radio based data network using commercial Wi-Fi gear with modified firmware.**

RADIO DESIGN

The presence of a signal is determined by the receiver's sensitivity.   A receiver's selectivity is the ability to discriminate between signals. These two goals conflict as the first part of the receiver is designed to be sensitive to a wide range of frequencies (sensitivity). Then, you need to narrow what you hear (selectivity).

To accomplish selectivity, the radio converts the incoming signal to an intermediate frequency using a mixer. Then, filtering stages operate on that one intermediate frequency and provide selectivity.

It all starts with an **oscillator – the circuit that generates a signal at a specific frequency.**

Then, you **convert a signal from one frequency to another with a mixer.**

You convert the RF input and output to another band using a transverter.   A transverter is a type of mixer. It is a "transverter" because it also transmits. The keyword is "output." That tells us you are converting a transmitter, thus a transverter. *Cheat: If you see transverter in an answer, it is always right.*

Power line noise or ignition noise might go away if you turn on the noise blanker. *Hint: Blank the noise with a noise blanker.*

Another receiver circuit is called **AGC – Automatic Gain Control keeps the received audio relatively constant**. AGC knocks down the gain if you come across a loud station so you don't blow out your ears.

If your receiver needs a little help with weak signals, **you would put an RF preamplifier between the**

antenna and receiver. It is an amplifier that is "pre" or before the radio and boosts the signal.

A mobile transceiver typically requires about 12 volts. You hook to the car battery. At home, you need a power supply to convert 120-volt house service. **The component commonly used to change 120V AC house current to a lower voltage is a transformer.** *Hint: A transformer transforms.* **The circuit that controls the amount of voltage from a power supply is a regulator.** *Hint: It regulates.*

Your radio runs off direct current, but the house power is alternating current. **The device that changes an alternating current to direct current is a rectifier.**

To determine the minimum current capacity needed for a transceiver power supply, you need to consider:
Efficiency of the transmitter at full output
Receiver and control circuit power
Power supply regulation and heat dissipation
All of the choices are correct.
All these factors affect how much power, and therefore the minimum current capacity, your station requires to operate.

Wiring between the power source and the radio should be heavy-gauge and kept as short as possible to avoid voltage falling below that needed for proper operation. All wire has some resistance and that resistance in series with the power supply can cause the voltage to drop. Don't use cheap thin speaker wire to power your radio.

Don't power mobile equipment from a cigarette lighter outlet. It may not be able to supply enough current and will blow a fuse or overheat the connectors. Also,

RADIO DESIGN

you may end up tied to a circuit that generates noise. See the next section, "Clean Up The Signal."

A stout and well-regulated power source will handle the load with some room to spare. Keep this in mind when you buy a power supply for the home shack. A robust supply designed to provide 35 amps will only cost about $30 more than a minimal 23 amp version, but you'll never have to worry about power when you get more radios or accessories. I like supplies with volt and amp meters so I can see what is going on.

The power supply is not glamorous but it is the foundation of the station. A good one will last a lifetime so, don't try to save a few dollars or you will regret it.

Stow your power supply away from the radio, so noise and hum from the supply doesn't couple to your receive or transmit signal. I keep mine under the desk.

CLEAN UP THE SIGNAL

You'll know a bad signal when you hear it. You can tell from the chirpy, raspy sounding CW or over-driven, over-processed or garbled audio.

Too high a microphone gain and the output signal may become distorted. It is the same as shouting into the microphone. Don't overdrive!

On FM, the amount of deviation is determined by the amplitude (loudness). **If you are told you are over-deviating, move away from the microphone.** You are too loud and your signal is too wide.

If you are told your transmissions are breaking up on voice peaks, you are talking too loudly. You are over-deviating, over-driving your transmitter.

A high-pitched whine in a mobile transceiver that varies with speed is usually the car's alternator.

Since the noise is in your transmitter, you won't hear it, but the person you're talking to might comment. The cure is to connect your radio directly to the battery. Two things happen. First, you have less chance of being tied into a circuit with noise. Second, the battery acts as a huge filter capacitor.

There is nothing wrong with asking for or giving a critical signal report. Everyone wants to sound good and no one wants to be a "lid."[12]

Sometimes, RF can get into the audio and be rebroadcast. That is RF feedback. **A symptom of RF feedback is reports of garbled, distorted or**

---

[12] "Lid" is an old telegraph expression referring to a poor operator. It is still used.

unintelligible transmissions. Poor grounding or being close to your antenna can cause feedback.

Shielded wire prevents coupling of unwanted signals to or from the wire. *Hint: A shield prevents coupling.*

RF can get into your microphone cable, power cable or cables connected to the computer. **To cure distorted audio caused by RF current on the shield of a microphone cable use a ferrite choke.** Ferrite is a mixture of ceramic and metal that increases the effectiveness of a coil to resist the passage of RF. A few turns of your cable through a ferrite choke may solve your problem. *Hint: A ferrite choke chokes off the RF.*

To reduce or eliminate interference to a nearby telephone put an RF filter on the telephone. The phone line is acting as an antenna, and you need to choke off the RF with an RF filter. *Hint: Filter RF with a filter.*

The first step to resolve cable TV interference is to be sure all the connectors are installed properly. Loose connectors break the shield and allow interference to enter or exit. Tighten them down and make sure any cable that is not connected has a terminating plug. That is a small cap that screws on the unused end of a TV cable.

Radio frequency interference may be caused by
Fundamental overload
Harmonics
Spurious emissions
All choices are correct
Hint: Lots of things cause interference. All are correct

An AM/FM radio might receive amateur radio transmissions because the receiver is unable to

reject strong signals outside the AM or FM band.
That is an example of fundamental overload. The
amateur radio signal is so strong; it overwhelms the
AM/FM radio.

**To reduce the overload of a non-amateur radio
or TV, block the amateur signal with a filter on
the antenna input of the receiver.** *Hint: The
question asks about overload on a non-amateur radio
or TV. That is a receiver, so the filter needs to go on
the input of the receiver to block the amateur signal.*

**A band-reject filter can reduce the overload to a
VHF transceiver from a nearby FM broadcast
station.** *Hint: The filter is going to reject the FM
band, same principal as above.*

**If a neighbor complains you are interfering with
their radio or TV, make sure your station is not
causing interference to your own radio or
television.** First, check your home. If you are clean
at home, the problem is at your neighbor's house.
Once they see your antenna, you become a target.
Hams get blamed even when not transmitting.

**Part 15 devices are unlicensed and may emit low
power radio signals on frequencies used by
licensed services.** Part 15 is the section of the FCC
regulations dealing with unlicensed devices.

Routers, cordless phones, weather stations, wireless
printers, audio and video equipment are all Part 15
devices. Read the label on one.

The FCC has authorized such devices with the
understanding their owner must accept interference
from a licensed service (you) but must not interfere
with a licensed service (you again). I don't know how
you tell the neighbor to get rid of his plasma TV (a

CLEAN UP THE SIGNAL

notorious Part 15 violator), but you are within your rights to do so.

If something in a neighbor's home is causing harmful interference to your amateur station,
Work with your neighbor
Check your station
Politely inform.
All the choices are correct.
Hint: There are lots of suggestions, so all the choices are correct.

Some wall wart power supplies, those little black boxes you plug in the wall to power electronics, can be a source of electronic noise. Also, watch out for fish tank heaters and motor controllers in dishwashers, washing machines and furnaces. If you experience an issue, run your radio on battery power and turn off circuit breakers in your house until the noise goes away. Then, you will have isolated the circuit with the problem and know where to look for the offending device. If you are not able to "turn off" the noise, it may be coming from a neighbor's house.

ELECTRONIC THEORY

Let's talk about some of the parts we "rescued" out of the old chassis behind the TV repair store.

The electrical part that **opposes the flow of current in a DC circuit is called a resistor.** Resistance is measured in ohms. Resistors have color-coded bands that tell their value.

Adjustable volume controls are potentiometers, variable resistors.

Inductors are composed of a coil of wire. Inductors store energy in a magnetic field.

The ability to store energy in a magnetic field is called inductance.

The basic unit of inductance is the henry. The more inductance, the more henries. *Remember "magnetic field, coils, henries, and inductance" go together.*

Impedance is the measure of opposition to AC current flow. The higher the frequency, the more a given henry impedes the flow. **The impedance is measured in ohms** (just like DC). If you know the value in henries, you can calculate the amount of impedance for a given frequency. We'll leave that math for the Extra Class license class test.

Two or more conductive surfaces separated by an insulator, make up a capacitor. The ability to store energy in an electrical field is called capacitance.

The electrical component that stores energy in an electrical field is called a capacitor.

ELECTRONIC THEORY

The basic unit of capacitance is called a farad. *Remember "electrical field, farads, and capacitors" go together.*

The higher the frequency, the less impedance a capacitor provides. It is the opposite of the inductor.

An inductor is combined with a capacitor to make a tuned circuit. A resonant or tuned circuit is an inductor and capacitor connected in series or parallel to form a filter.

If an ohmmeter indicates low resistance and then increasing resistance with time, the circuit contains a large capacitor. It is charging up and, as it gets full, the resistance increases.

When a power supply is turned off, you might still receive a shock from stored charge in a large capacitor. Power supplies often use large capacitors as filters.

Capacitors hold a charge. We'd get a big capacitor out of a TV power supply, charge it up and then toss the electrostatic bomb to some poor and unsuspecting classmate. Of course, he would catch it, and get a nasty jolt. We were way beyond handshake-buzzer mischief. If you did now what we did then, you would be arrested.

Those old TVs ran on vacuum tubes, but we did plenty of experimenting with transistors. A transistor is sometimes called a semiconductor. **A component made of three layers of semiconductor material is a transistor.**

Transistors are components capable of using a voltage or current to control current flow.

A transistor can be used as a switch or an amplifier. Electronic components that can amplify signals are transistors.

The term that describes a transistor's ability to amplify a signal is called gain. The primary gain-producing component in an amplifier is a transistor.

An FET is a special kind of transistor called a Field Effect Transistor.

The component that allows current to flow in only one direction is a diode. Diodes are also called rectifiers. **The device or circuit that changes an alternating current to a direct current is a rectifier.** That is their job in a power supply.

The two electrodes of a diode are called the anode and cathode. The anode is the positive side, and the cathode is negative. *Hint: Remember "Annie and Cathy."*

The cathode end of a diode is often marked with a stripe. Diodes have polarity (a plus and a minus side), and must be installed in the right direction.

The abbreviation LED stands for light emitting diode. An LED is commonly used as a visual indicator. LEDs can be in flashlights or indicator lights on your equipment. They produce a very bright light while using little power and last forever.

VOLTS, OHMS, AMPERES & POWER – OHM'S LAW

Ohm's Law describes the relationship between voltage, amperage and resistance. The math involved is just multiplication or division, and is certainly within the grasp of any high school student. Here is how you sort them out.

Voltage represents electromotive force. The unit of electromotive force is the volt. Think of Voltage as the pressure in the line. All that force can just sit there as it does in an electrical outlet. With nothing plugged in, the 120 volts are there, but not doing anything.

A circuit where the voltage is the same across all components is in parallel. Components that are side-by-side will all have the same pressure (voltage). *Hint: That is why all your household electric outlets have the same voltage.*

At the junction of two components in parallel, the current divides depending on the value of the components. The voltage remains the same but the current divides. **The voltage across each of two components in parallel is the same as the voltage at the source.**

Amperes represent the current or amount of electricity flowing through the circuit. When you turn on a light, current flows through the wire and powers the bulb. Now, that voltage is pushing the current through the load (light bulb).

Components in series (end-to-end) will have the same current through each of them. The current flow is the same in each one. *Hint: If you connect a* *small hose to a large hose, the same amount of water flows, but the pressure changes.*

A circuit where the current is the same through all the components is in series. At the junction of two components in series, the current remains the same. The same current still has to flow through the circuit. The voltage (pressure) changes through each component. **The voltage across each component is determined by the type and value of the components.**

Power is the amount of work that can be done by the pressure and flow.

There is a relation among the four:
More pressure = more current flow.
More resistance = less current flow.
More current or more pressure = more power.
Can this relation be expressed mathematically? Georg[13] Ohm thought so and came up with his famous formula known as Ohm's Law.

Ohm's Law: Voltage = Amperes times Resistance. For this amazing feat, he got to name the unit of resistance after himself (ohm).

E symbolizes voltage (electromotive force).
I is Amps or Amperage (flow).
R is Resistance.
Memorize that: Voltag**E**, **R**esistance, and the other one, Amps, **I.**

So the formula for Ohm's Law looks like E=IR.
If you know two of the variables, you can calculate the third. The easy way is to use the magic circle and draw it as the Eagle flies over the Indian and the Rock.

---

[13] That is not a typo. Georg had no "e" at the end of his name.

OHM'S LAW

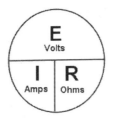

Put your thumb on the value you are solving, and the formula is what's left. For example, if you are solving for R, cover up the R <u>and</u> the answer is E/I, <u>voltage</u> divided by amperage.   If you are solving for E, cover it up, and the answer is I x R, amps times ohms. Try these out:

The formula used to calculate current in a circuit is voltage divided by resistance. *Hint: It asks for current, so cover the "I," and the result is E divided by R.*

The formula used to calculate voltage in a circuit is current multiplied by resistance. *Hint: Cover the "E," and the result is I time R.*

The formula used to calculate resistance in a circuit is voltage divided by current. *Hint: Cover the "R," and the result is E divided by I.*

The pool questions using Ohm's Law are in the Quick Summary starting at question T5D04. They are variations on the same question. You will get one on your test. It should be easy if you keep calm, use the magic circle and watch your decimal point. Go there now and solve some for practice.

Hint: When you take your test, the first thing you do is draw the magic circle on the back of the test sheet or scratch paper as allowed by the VE team. Don't write in the test book as that is used again. When the Ohm's Law question comes up, refer to your sketch.

POWER IN WATTS

The rate at which electrical energy is used is called power. Electrical power is measured in Watts.

The formula to calculate electrical power in a DC circuit is Power equals Voltage (E) multiplied by Current (I). It is easy as PIE, P=IE.

We have another magic circle to help us. *Hint: Copy it to the back of the test sheet as well.*

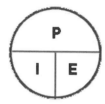

Cover the answer and solve using the two known amounts.

There are three possible test questions starting at T5C08 in the Quick Summary. Go there now and solve them for practice.

HOW TO DRAW A RADIO

The plans for the guts of electronic equipment are drawn out in a schematic diagram. Schematic diagrams are road maps that show how the components are connected.

Schematic symbols represent electrical components. The symbol is often a fair representation of the component's design or function which helps you identify it.

The question pool has three schematic diagrams. They are at the back of this book, before the Index. You will be given only one diagram as part of your test package. The test asks you to identify some of the components. The only components you need to identify are in the following questions. Don't fret over the other components.

Look at Figure T-1, in the back of the book before the Index.

Component 1 is a resistor. *Hint: Imagine the resistance if you had to run a zigzag line like that.*

Component 2 is a transistor. The function of the transistor is **to control the flow of current. A transistor can be an amplifier or a switch.**

A switch connects or disconnects circuits.
The transistor looks like it is a switch in this circuit. A small change in the voltage applied through the resistor will cause the transistor to conduct and light the lamp.

Component 3, a lamp and looks like a lamp.

Component 4 is a battery with long and short lines representing the cells. If there were only two lines, you would be looking at a capacitor.

In Figure T-2, **Component 3 is a switch.** The one in the diagram is **single-pole single-throw.** *Hint: One wire in and one wire out*

Component 4 is a transformer. See the sets of windings (coils). The two coils couple together through mutual inductance. One side induces current and voltage on the other side. The ratio of the number of windings determines the step-up or step-down result.

A transformer is used to change 120v AC house current to lower voltage for other uses. The little cube you plug in to charge your cell phone is a power supply with a step-down transformer.

Component 6 is a capacitor. A capacitor has two sides, and so does its symbol. If there are more than two lines in the drawing, you are looking at the symbol for a battery.

Component 8 is a light emitting diode. *Hint: The arrows are light coming out of the LED.* **A light emitting diode is commonly used as a visual indicator.** The buttons that light up on your TV remote are LEDs. **LED means light emitting diode.**

Component 9 is a variable resistor. The zigzag line is a resistor; the arrow makes it variable.

Look at Figure T-3.

Component 3 is a variable inductor. An inductor is a coil. You can change the value of a variable inductor. One method is by tapping into a different

HOW TO DRAW A RADIO

turn. *Hint: The symbol looks like a coil. Imagine the arrow sliding up and down to change the value.*

Component 4 is an antenna. *Hint: It looks like a funnel to collect radio waves.*

A tuned circuit is composed of an inductor and a capacitor. In the case of figure T-3, we have two variable capacitors and a coil to form a tuned circuit. **A resonant or tuned circuit is an inductor and capacitor connected in series or parallel that form a filter.** *Cheat: Ditch the overly-complicated question. Just know an inductor and a capacitor together form a tuned circuit or filter.*

Some other components include:

The component that displays an electrical quantity as a numeric value is a meter.

A relay is an electrically controlled switch. For example, the ignition switch in your car could never handle the tremendous surge of power needed to start the car engine when you turn the key. A small amount of power supplied by the ignition switch closes a relay with big contacts that pass power from the battery to the starter motor.

The type of circuit that controls the voltage from a power supply is called a regulator. *Hint: A regulator regulates.*

The device that combines several semiconductors and other components into one package is an integrated circuit. It integrates (combines) several circuits on one component, sometimes called a "chip."

Page 86   Technician Class – The Easy Way

DECIBELS

Decibels are logarithmic (Power of 10) and relate to relative change. **Going from 20 to 200 watts is 10 times or 10 dBs.** Going from 200 watts to 2000 watts is another 10 db.

Since we are talking about logarithms, the change is not linear. **Going from 5 watts to 10 watts is 3 dB**. Doubling is 3 dB. It doesn't matter where you start, if you double, it is 3 dB, and if you double again, it is 6 dB.

If you halve the power it is -3 dB. Halve the power again and it is -6 dB. **A power decrease from 12 watts to 3 watts is -6 dB.** Minus half, and minus half again.

An example of using decibels is to compare the results of a better antenna versus an amplifier. A good 2 meter directional antenna might have a gain of 9 dB. Each 3 dB is a double, so that is double, double and double = 8 times the power. To get the same result for your 50-watt desktop radio, you would have to buy a 400-watt amplifier. The antenna is a lot cheaper and will give you gain on the receiver side as well.

Decibels can also be used to compare different types of coax. The loss is measured in decibels per hundred feet at different frequencies. If you lose 3 dB, you have lost half your power. On 2 meters, RG58 coax loses 5.1 dB per hundred feet and only 31 out of a hundred watts makes it to the antenna. RG8 only loses 1.8 dB and 63 watts makes the trip.

MOVING DECIMALS

A kilovolt would be one-thousand volts. Kilo is thousand. Mega is million. Giga is thousand million.

Another way to specify a frequency of 1,500,000 hertz is 1500 kHz. Kilo means thousand so 1500 KHz is 1500 thousand hertz or 1,500,000 hertz or 1.5 Megahertz (MHz). They all say the same thing.

2425 MHz would be the same thing as 2.425 GHz. Gigahertz. A Gigahertz is a thousand megahertz.

28,400 kHz would be the same as 28.400 MHz. *Hint: When going to a larger identifier (kHz to MHz), put in a comma and then change it to a period.*

A frequency display of 3.525 MHz is the same as 3,525 kHz. *Hint: When going to a smaller identifier (MHz to kHz) change the period to a comma.*

For really small numbers we use, milli (1/1000) and micro (1/1,000,000.) and pico (1/1,000,000,000,000).

One microvolt is one, one-millionth of a volt.

A 3000 milliampere current would be = 3 amperes. 3000 /1000 *Hint: Put a comma in 3000 and change it to a period.*

500 milliwatts would be 0.5 watts 500/1000

1.5 amperes would be 1500 milliamperes. *Hint: How many thousandths? 1.5 X 1000*

Picofarads are tiny: one-millionth of one-millionth of a farad. So **1,000,000 picofarads =** one-millionth of a farad or **1 microfarad**. *Cheat: Memorize it.*

BUILDING EQUIPMENT AND MEASURING VALUES

Radio and electronic soldering use rosin-core solder. Acid-core plumber's solder would eat away the connection.

A cold solder joint is grainy and dull. Heat the joint and flow the solder onto the connection. If you see a cold solder joint, it means there was not enough heat or the part moved before the solder set. Reheat the connection. Cold solder joints are poor mechanical and electrical connections.

Multimeters measure voltage and resistance. They are very handy and some cost under $10. The cheap ones are not laboratory grade, but they will give accurate enough readings for most uses.

Voltage is electromotive force. The basic unit is the volt. **We could measure it with a voltmeter hooked up in parallel with (across) the circuit –** from one end of the battery to the other.

We **measure amperes (amps) with an ammeter. An ammeter is connected in series (in line) with the circuit.** You measure flow through the circuit.

If you're measuring resistance, be sure the circuit is not powered. A variation on the same question: **You'll damage the meter if you try to measure voltage when using the resistance setting.**

The resistance setting is very delicate and running voltage through it will pin the needle or let the smoke out of your meter. That is a ham-speak meaning you fried it. If you are lucky, your meter had a fuse that blew before being damaged.

ANTENNAS –
PUT SOME FIRE IN YOUR WIRE

POLARIZATION

Radiation from a half-wave dipole antenna is strongest broadside to the antenna. Most of the energy will be broadside to the wire and not off the ends.

The orientation of the electrical field describes a radio wave's polarization. *Hint: Remember "electrical" field determines polarization.*

A simple dipole oriented parallel to the Earth's surface is horizontally polarized. The wire is horizontal, and the radiation is strongest broadside, so the radiation is also horizontal. **The antenna polarization used for long distance CW and SSB on the VHF and UHF bands is usually horizontal.** *Cheat: You are reaching out beyond the horizon.*

A vertical antenna has an electrical field perpendicular to the Earth. Vertical antennas also radiate broadside, and since the antenna is vertical, they are vertically polarized. Mobile FM uses vertical polarization because that is the way an antenna fits on a car and the way you hold an HT: antenna up-and-down. Therefore repeaters also use vertical antennas.

If the antennas on opposite ends of a VHF or UHF lie are not kept in the same polarization, signals could be significantly weaker. Don't hold your rubber ducky sideways as that will make the antenna horizontally polarized. If you are talking on a vertically polarized repeater, your signal could be as much as 100 times weaker.

When a signal bounces off something, it becomes wobbly like a tumbling hula hoop. It becomes elliptically polarized (random both ways), and it doesn't matter how your antenna is set up. **Skip signals are elliptically polarized, and you can use either a vertically or horizontally polarized antenna**.

Irregular fading can be caused by the random combining of signals arriving from different paths. Signals arriving at slightly different times or different polarizations because they bounced off different objects are multi-path. **If data signals arrive via multiple paths, error rates are likely to increase.** The receiving station gets confused.

ANTENNA LENGTHS

I told the salesman I needed some antenna wire. He asked me, "How long do you need it?" I thought that was an odd question so I answered, "I'm building an antenna, so I guess I'll need it for a long time." He coax-ed me out the door.

How long should an antenna be? The classic and most fundamental antenna is one-half wavelength long and fed in the middle with coax. If you think about the current flowing in the wire, it will start out high in the middle at the feed point, and by the time it gets to the end, the cycle will be ready to reverse. Nothing is left over, and nothing is reflected back.

A half-wavelength antenna is half a wavelength long. The test question asks, **"How long is a 6 meter half-wavelength dipole?"** Six meters is a full wavelength, so 6 meters / 2 = 3 meters long for a half wave. That is a little over 9 feet x 12 = **112 inches is the closest answer.**

ANTENNAS

Many verticals are quarter-wave, and the ground or the vehicle body provide the other half of the antenna.

A quarter-wave for 146 MHz is 19 inches. 146MHZ is the 2 meter band. One quarter of 2 meters is one-half meter. A meter is about 36 inches, so one half would be about 19 inches. Closest answer.

To make an antenna resonant at a higher frequency, you would shorten it. Higher frequency means shorter wavelength because the wave can't travel as far before it reverses. If your antenna analyzer shows the antenna resonant at a lower frequency than you expected, shorten it to raise the resonant point.

The advantage of a 5/8 wavelength antenna is that it has a lower angle of radiation and more gain than a 1/4 wave. The slightly longer length squashes the signal pattern down closer to the horizon. You will see 5/8 wave antennas offered for car installation.

If your calculations tell you the antenna will to be too long for your installation (an 80-meter dipole is about 135 feet long), you can add an inductor to shorten the wire. **A type of loading can be inserting an inductor to make an antenna electrically longer.** You won't have this problem on VHF/UHF because the antenna sizes are quite manageable.

DIRECTIONAL ANTENNAS

Yagis, Quads, and Dishes are all directional antennas. They concentrate the signal in one direction and reject signals to the back and side of the antenna. **A beam antenna concentrates signals in one direction.** We call them "beams" because they beam the signal in one direction. A Yagi is a type of beam.

The gain of an antenna is **the increase in signal strength in a specific direction compared to a reference antenna.** *Hint: Too complicated, just remember gain increases strength for both antennas and amplifiers.*

A directional antenna has gain on both transmit and receive. That is an important consideration. In our discussion of decibels, we found 9 dB gain from a beam antenna was the same as adding a 400-watt amplifier. Antennas are reciprocal meaning the antenna also gives 9 dB of gain on receive but the amplifier does not. Spend your money on antennas before amplifiers.

To locate the source of noise interference or jamming use radio direction finding. Use a directional antenna to get a bearing on the signal from two locations. The target is where the bearings cross on a map.

A directional antenna is also useful for a hidden transmitter hunt. This sport is also called fox-hunting. There are national and international fox-hunting competitions, but the fox doesn't get hurt.

When using a directional antenna, you might be able to access a distant repeater if buildings or obstructions are blocking the direct line of sight path by finding a path that reflects signals to the repeater. *Hint: You point your antenna to bounce around the obstruction.*

COAXIAL CABLES

The line carrying RF to the antenna is called feed line. Coaxial cable is a type of feed line that has a center conductor, a layer of insulation, and then a braided shield all covered with a tough outer cover.

ANTENNAS

Coax is favored over other feed line because coax is easy to use and requires few special installation considerations. Coax is very versatile. You can run it anywhere including next to metal such as aluminum siding. If you have extra, coil it up on the floor. You can install coax outdoors or indoors, and bury certain kinds. *Remember, coax is convenient.*

The characteristic impedance most often used for coax is 50 Ohms. It matches the normal impedance of a dipole installed over the ground.

When increasing frequency, coaxial cable loss increases. Higher frequency equals higher loss. When choosing coax, check the loss attenuation, measured in dB per hundred feet at various frequencies. You don't want to lose your signal in the feed line. A 6 dB loss means only one-quarter is getting through. The rest is converted to heat and lost!

Coaxial cable comes in many flavors. RG-58 and RG-8 are two popular types. **The electrical difference between RG-58 and RG-8 coaxial cable is the RG-8 has less loss.** Larger RG-8 cable will have less loss than smaller RG-58.

The lowest loss at VHF and UHF would be air-insulated hard line. It is "hard line" because it is very rigid. That makes it hard to bend. **The disadvantage of air core coaxial cable is that it requires special techniques to avoid water absorption.** Water is the enemy of coax. Commercial installations use hard line but most hams avoid it.

The most common cause for the failure of coaxial cable is moisture contamination. Water gets in, and the braided shield acts as a wick. The leak can come from the connectors or wildlife. Squirrels love to chew on coax, and they gnaw through the outer jacket exposing the braid. I inspect my coax every couple of years, and every time the antenna comes down. I always find some squirrel damage.

A PL-259 connector is commonly used on HF frequencies. The PL-259 is a screw-on plug that has been around since the 1930's. It is not waterproof but is relatively easy to solder to coax cable.

Frequencies above 400MHz use a Type N connector. The type N is a different plug design that is waterproof. It is harder to solder to coax.

Seal the connectors to prevent an increase in feed line loss (from moisture). Minimally, use good quality electrical tape rated for UV (sun) exposure.

Coax outer jacket should be resistant to ultraviolet light because ultraviolet light from the sun can damage the jacket and allow water to enter. Baking in the sun for years can destroy the jacket causing it to crack or peel. Good quality coax from name-brand manufacturers can last for decades if the squirrels don't chew through it first. It is worth inspecting your coax for damage every couple of years.

ANTENNA ANALYZERS AND SWR

When we weren't busy trying to make gunpowder or some other mischief, we would put up antennas. I don't think we knew what an antenna analyzer was, and the concept of SWR was also a bit foreign. We just followed the formulas and hung wire in the air. For the most part, it worked and we made contacts. What you don't know can't hurt you.

ANTENNAS

Use an antenna analyzer to determine if the antenna is resonant. By resonant, we mean how well the antenna matches to the frequency. **You can also measure proper match with a directional wattmeter.** It shows the forward and reflected power. You want to minimize the reflected power.

The standing wave ratio (SWR) is a measure of how well the load is matched. Reflected power is compared against forward power. Obviously, less reflected power is better. **You want a low SWR in coaxial line to reduce signal loss.**

Power lost in the feed line is converted to heat. You want signal, not heat.

The proper location for an SWR meter is in series with the feedline between the transmitter and the antenna. It is measuring the forward and reverse power coming out of and into the transmitter and antenna.

A perfect impedance match is an SWR of 1 to 1. Don't get too crazy chasing the elusive 1:1. Often, that is just not achievable, and the difference in signal is not significant. Most of us would be satisfied with 1.5 to 1 and aren't concerned until the radio starts to reduce power.

Most radios will automatically reduce power as SWR increases to protect the output amplifier transistors. The radio has a built-in defense mechanism.

An SWR of 4:1 would indicate an impedance mismatch.

You can use an antenna tuner (antenna coupler) to help match the antenna system impedance to the transceiver. An antenna tuner is a combination of variable capacitors and inductors to make up for the mismatch. If you have a mismatch, it is best fixed at the antenna. Applying a band-aid on the transmitter won't stop the losses that occur in coax when the SWR is too high.

If you have erratic changes in SWR readings, you might have a loose connection in the antenna or feed line. Erratic changes would indicate an intermittent connection, a broken wire or a faulty connector.

Make a chart of your SWR readings so you can compare over time and see if anything is changing in your system. If your SWR starts climbing, you probably have moisture in the coax. The loss of signal strength will be so gradual you may not hear the change.

A dummy load prevents transmission of signals over the air when making tests. We use a dummy load when we don't want our signal to go out over the air.

A dummy load consists of a non-inductive resistor and heat sink. The resistor absorbs the transmitter current, and the heat sink dissipates the heat generated by power burned up in the resistor.

BATTERIES

Rechargeable batteries include
Nickel-metal hydride
Lithium-ion
Lead-acid gel cell.
All choices are correct.
Hint: There are many kinds of rechargeable batteries,
so all are correct.

Carbon Zinc are non-rechargeable batteries.

A mobile transceiver requires about 12 volts.
Mobile transceivers use the same voltage as a car
battery which is a nominal 12 volts.

You can recharge a 12-volt lead-acid battery if the
power is out by connecting it in parallel with a
vehicle's battery and running the engine. A car can
run on a full tank of gas for about 72 hours. If you
ran the car a few hours a day to charge batteries, it
could last for weeks. Consider your car as a source of
energy in case of a power outage.

If you try to charge or discharge a lead-acid
battery too quickly, it can overheat and give off
flammable gas or explode.

A safety hazard of 12-volt storage batteries is
that shorting the terminals can cause burns, fire
or an explosion. Lead acid batteries deliver a
tremendous amount of current that can melt a ring or
tool. Take your ring or metal watchband off and keep
your tools clear when working around car batteries.

Connect the negative side of the power cable to
the battery or engine block ground strap. The
negative is the black wire, and it is considered the
"ground."

*Congratulations! You're done. Now run through the
Quick Summary and, when you feel confident you
know the answers, try some practice tests online. Go
to QRZ.com and click on the Resources tab. Good
luck.*

73/DX – Buck, K4IA

Here's my
HF/VHF/UHF mobile
rig installed in a
Jaguar XJ6. The
radio is in the
trunk. A control
head is in what
used to be the
ashtray. The Morse
code paddles are to
the left of the gear
shift.

This is what I call
"the cat's meow."

TECHNICIAN CLASS

QUICK SUMMARY

In this section, the questions are in regular typeface, and the answers are in bold. The questions and answers have been distilled down to the minimum information needed. Learn to recognize the answer.

The material is grouped by the subelement (T1) and group (T1A). You will have one question from each group.

SUBELEMENT T1 – FCC Rules, descriptions, and definitions for the Amateur Radio Service, operator and station license responsibilities –
[6 Exam Questions - 6 Groups]

T1A01 A purpose of the Amateur Radio Service is **advancing skills in the technical and communication phases of the radio art.**

T1A02 The agency regulating and enforcing the rules in the United Staes is the **FCC.**

T1A03   The FCC **encourages** the use of a phonetic alphabet.

T1A04 One person may only hold **one** operator/primary station license grant.

T1A05 Proof of possession of an FCC-issued license is **appearing in the FCC database.**

T1A06 The FCC definition of a "beacon" is transmitting communications for the purposes of **observing propagation or related experimental activities.**

T1A07 The FCC Part 97 definition of a "space station is an amateur station **located more than 50 km above the Earth's surface.**

T1A08 The entity that recommends transmit/receive channels for auxiliary and repeater stations is a **Volunteer Frequency Coordinator recognized by local amateurs.**

T1A09 A Frequency Coordinator is selected by **local amateur operators**.

T1A10 The Radio Amateur Civil Emergency Service (RACES)is a radio service using **amateur stations, frequencies, and operators for civil defense. All the choices are correct.**

T1A11 Willful interference to other amateur radio stations is permitted **at no time.**

T1B01[14] The International Telecommunications Union (ITU) is a **United Nations agency**.

T1B02 **Any amateur holding a Technician or higher-class license** may make contact with the International Space Station.

T1B03 A frequency is within the 6 meter amateur band is **52.525 MHz.**

T1B04 If you are transmitting on 146.52 MHz, you are in the **2 meter band.**

T1B05 Emissions on the frequencies between 219 and 220 MHz are limited to **fixed digital message forwarding systems only.**

T1B06 A Technician class operator has HF phone privileges on **10 meters only**.

---

[14] Note we have moved to Group T1B, and you will get one question from this group. Groups are separated by ~~~~~~~~~~.

T1B07 The VHF/UHF frequency ranges limited to CW only are **50.0 MHz to 50.1 MHz and 144.0 MHz to 144.1 MHz.** *Cheat: This is the only answer with VHF frequencies.*

T1B08 If the Amateur service is secondary, U.S. amateurs **may find non-amateur stations in those portions and must avoid interfering with them.**

T1B09 You should not set your transmit frequency to be exactly at the edge of an amateur band or sub-band
To allow for calibration error in the frequency display
So that sidebands do not extend beyond the band edge
To allow for transmitter frequency drift
All of these choices are correct,
Hint: Lots of reasons to stay away from the edge.

T1B10 The Technician class operator has HF privileges for RTTY and data transmissions for **10 meters only.** *Hint: HF Phone, CW and RTTY only on 10 meters.*

T1B11 The maximum peak envelope power output for Technician class operators on the HF bands is **200 watts.**

T1B12 The maximum peak envelope power output for Technician class operators using frequencies above 30 MHz is **1500 watts.** *Hint: Above 30 MHz is no longer HF; it is VHF/UHF.*

~~~~~~

T1C01 License classes are **Technician, General, Amateur Extra.**

T1C02 **Any licensed amateur** may select a desired call sign under the vanity call sign rules.

T1C03 An FCC-licensed amateur radio station is permitted to make international communications **incidental to the purposes of the Amateur Radio Service and remarks of a personal character.**

T1C04 You are allowed to operate your amateur station in a foreign country **when the foreign country authorizes it.**

T1C05 A valid call sign for a Technician class amateur radio station could be **K1XXX.**
Hint: Technicians can request a 1x3 call sign through the vanity licensing program. First you get your assigned call, then you request a change. Remember, 3 X.

T1C06 An FCC-licensed amateur station may transmit from **any vessel or craft located in international waters and documented or registered in the United States.**

T1C07 If correspondence from the FCC is returned as undeliverable **revocation or suspension may result.**

T1C08 The term for an FCC-issued license is **ten years.**

T1C09 The grace period following the expiration within which the license may be renewed is **two years.**

T1C10 You may operate a transmitter as **soon as your operator/station license grant appears in the FCC's license database.**

T1C11 If your license has expired and is still within the allowable grace period, **you may not continue to operate** a transmitter until the FCC license database shows that the license has been renewed.

~~~~~~

T1D01 FCC-licensed amateur radio stations are prohibited from exchanging communications with any country **that objects to such communications.**

T1D02 One-way transmissions are allowed **when transmitting code practice, information bulletins, or transmissions necessary to provide emergency communications.** *Hint: "Emergency communications."*

SUMMARY – RULES

T1D03 Messages encoded to hide their meaning are allowed **only when transmitting control commands to space stations or radio control craft.**

T1D04 An amateur station is authorized to transmit music **when incidental to manned spacecraft communications.**

T1D05 Amateur radio operators may use their stations to notify other amateurs of equipment for sale or trade **when the equipment is normally used in an amateur station, and such activity is not conducted on a regular basis.**

T1D06 The restriction concerning transmission of indecent or obscene language **is such language is prohibited.** *Reminder: No one maintains a list.*

T1D07 The types of amateur stations that can automatically retransmit the signals of other amateur stations are **repeater, auxiliary, or space stations.** *Remember: RAS*

T1D08 A control operator of an amateur station may receive compensation for operating **when the communication is incidental to classroom instruction at an educational institution.**

T1D09 Amateur stations are authorized to transmit signals related to broadcasting, program production, or news gathering, only where **directly relate to the immediate safety of human life or protection of property.**

T1D10 "Broadcasting" means **transmissions intended for reception by the general public.**

T1D11 An amateur station may transmit without on-the-air identification **when transmitting signals to control model craft.**

T1E01 An amateur station is **never** permitted to transmit without a control operator.

T1E02 Any amateur **whose license privileges allow them to transmit on the satellite uplink frequency** may be the control operator of a station communicating through an amateur satellite or space station. *Hint: You transmit on the uplink frequency so, you need privileges there.*

T1E03 The **station licensee** designates the station control operator.

T1E04 The transmitting privileges of an amateur station are determined by **the class of operator license held by the control operator.**

T1E05 An amateur station control point is **the location at which the control operator function is performed.**

T1E06 A Technician class licensee may be the control operator of a station operating in an exclusive Amateur Extra class operator segment of the amateur bands **at no time.**

T1E07 If the control operator is not the station licensee, the **control operator and the station licensee are equally responsible** for the operation of the station.

T1E08 An example of automatic control is **repeater operation.**

T1E09 During remote control operation:
The control operator must be at the control point
A control operator is required at all times
The control operator indirectly manipulates the controls
All these choices are correct.
Hint: Two out of three makes all correct.

T1E10 An example of remote control is **operating the station over the internet.**

SUMMARY – RULES

T1E11 The FCC presumes the control operator is **the station licensee.**

~~~~~~

T1F01 The station licensee must make the station and its records available for FCC inspection **at any time upon request by an FCC representative.**

T1F02 When using tactical identifiers such as "Race Headquarters" you must transmit the station's FCC-assigned call sign **at the end of each communication and every ten minutes during a communication.** *Remember "every ten minutes."*

T1F03 An amateur station is required to transmit its assigned call sign at least **every 10 minutes during and at the end of a communication.** *Same as above.*

T1F04 An acceptable language to use for station identification is the **English language.**

T1F05 Call sign identification for a station transmitting phone signals can be **CW or phone emission.** *You'll hear voice repeaters identifying in CW.*

T1F06 An acceptable self-assigned indicator when identifying using a phone transmission is
KL7CC stroke W3
KL7CC slant W3
KL7CC slash W3
All of these choices are correct.
Hint: You do not need to memorize the call signs, just know "stroke, slant, slash" mean the same thing.

T1F07 When a non-licensed person is allowed to speak to a foreign station, the foreign station must be one with **which the U.S. has a third-party agreement.** *Hint: non-licensed = third party.*

T1F08 The term "Third Party Communications" means **on behalf of another person.**

T1F09 A station that simultaneously retransmits the signal of another amateur station on a different channel or channels is a **repeater station.**

T1F10 If a repeater inadvertently retransmits communications that violate the FCC rules, **the control operator of the originating station is responsible.**

T1F11 A requirement for the issuance of a club station license grant is the **club must have at least four members.**

SUMMARY – OPERATING PROCEDURES

SUBELEMENT T2 - Operating Procedures –
[3 Exam Questions - 3 Groups]

T2A01 A common repeater frequency offset in the 2 meter band is **plus or minus 600 kHz.**

T2A02 The national calling frequency for FM simplex operations in the 2 meter band is **146.520 MHz.**

T2A03 A common repeater frequency offset in the 70 cm band is **plus or minus 5 Mhz**. *Hint: Wide band wide offset.*

T2A04 To call another station on a repeater if you know the other station's call sign, **say the station's call sign, then identify with your call sign.**

T2A05 To respond to a station calling CQ, **transmit the other station's call sign followed by your call sign.** *Hint: Call him first.*

T2A06 When making on-the-air test transmissions, **identify the transmitting station.**

.
T2A07 "Repeater offset" is the **difference between a repeater's transmit frequency and its receive frequency**.

T2A08 The procedural signal "CQ" means **calling any station**.

T2A09 The brief statement that indicates that you are listening on a repeater and looking for a contact **is your call sign.** *Hint: The briefest of all the answers.*

T2A10 A band plan, beyond the privileges established by the FCC is **a voluntary guideline** for using different modes or activities within an amateur band.

T2A11 The kind of communication taking place when an amateur station is transmitting and receiving on the same frequency is **simplex**.

T2A12 When choosing an operating frequency for calling CQ,
Listen first to be sure that no one else is using the frequency
Ask if the frequency is in use
Make sure you are in your assigned band
All of these choices are correct.

T2B01 The most common use of the "reverse split" function of a VHF/UHF transceiver is to **listen on a repeater's input frequency.**

T2B02 The sub-audible tone transmitted along with normal voice audio to open the squelch of a receiver is **CTCSS.**

T2B03 If a station is not strong enough to keep a repeater's receiver squelch open, you might receive the station's signal by **listening on the repeater input frequency.**

T2B04 If you are unable to access a repeater whose output you can hear, it could be:
Improper transceiver offset
The repeater may require a proper CTCSS tone
The repeater may require a proper DCS tone
All of these choices are correct.

T2B05 If a repeater user says your transmissions are breaking up on voice peaks, **you are talking too loudly.**

T2B06 The tones used to control repeaters linked by the Internet Relay Linking Project (IRLP) protocol are **DTMF**

T2B07 To join a digital repeater's "talk group," **program your radio with the group's ID or code.**

T2B08 When two stations transmitting on the same frequency interfere with each other, **common courtesy should prevail, but no one has an absolute right to an amateur frequency.**

SUMMARY – OPERATING PROCEDURES

T2B09 A "talk group" on a DMR digital repeater is a way for groups of users to **share a channel at different times without being heard by other users on the channel.**

T2B10 The Q signal that indicates you are receiving interference from other stations is **QRM.**

T2B11 The Q signal that indicates you are changing frequency is **QSY**.

T2B12 Simplex channels are designated in the VHF/UHF band plans so that **stations within mutual communications range can communicate without tying up a repeater.**

T2B13 SSB phone is used in amateur bands above 50 MHz in **at least some portion of all these bands.**

T2B14 A linked repeater network is a network of repeaters where **signals received by one repeater are repeated by all the repeaters.**

T2C01 When do FCC rules NOT apply to the operation of an amateur station? **Never, FCC rules always apply.**

T2C02 The term "NCS" used in net operation means **Net Control Station.** *Remember "net control."*

T2C03 When using voice modes to ensure that voice messages containing unusual words are received correctly, **spell the words using a standard phonetic alphabet.**

T2C04 RACES and ARES have in common that both organizations **may provide communications during emergencies.**

T2C05 The term "traffic" in net operation refers to **formal messages** exchanged by net stations.

T2C06 An accepted practice to get the immediate attention of a net control station when reporting an emergency is to **begin** your transmission by saying "Priority" or "Emergency" followed by your call sign.

T2C07 An accepted practice for an amateur operator who has checked into a net is to **remain on frequency without transmitting until asked to do so by the net control station.**

T2C08 Good traffic handling is **passing messages exactly as received.**

T2C09 Amateur station control operators are permitted to operate outside the frequency privileges of their license class only if necessary **in situations involving the immediate safety of human life or protection of property.**

T2C10 The information contained in the preamble of a formal traffic message is the **information needed to track the message.**

T2C11 The term "check," refers to the **number of words or word equivalents in the text portion of the message.**

T2C12 The Amateur Radio Emergency Service (ARES) is licensed amateurs who have **voluntarily registered** their qualifications and equipment for communications duty in the public service.

SUMMARY – RADIO WAVES, PROPAGATION

SUBELEMENT T3 – Radio wave characteristics: properties of radio waves; propagation modes –
[3 Exam Questions - 3 Groups]

T3A01 If another operator reports that your station's 2 meter signals were strong just a moment ago, but now they are weak or distorted, **try moving a few feet or changing the direction of your antenna if possible**, as reflections may be causing multi-path distortion.

T3A02 The range of VHF and UHF signals might be greater in the winter because of **less absorption by vegetation**.

T3A03 The antenna polarization normally used for long-distance weak-signal CW and SSB contacts using the VHF and UHF bands is **Horizontal**. *Cheat: Long distance is over the horizon.*

T3A04 If the antennas at opposite ends of a VHF or UHF line of sight radio link are not using the same polarization signals **could be significantly weaker.**

T3A05 When using a directional antenna, you might be able to access a distant repeater if buildings or obstructions are blocking the direct line of sight path, by **trying to find a path that reflects signals to the repeater.** *Point the antenna to bounce off an obstruction.*

T3A06 The rapid fluttering sound sometimes heard from mobile stations that are moving while transmitting is called **picket fencing.**

T3A07 The type of wave that carries radio signals between transmitting and receiving stations is called **Electromagnetic.**

T3A08 A likely cause of irregular fading of signals received by ionospheric reflection is **random combining of signals arriving via different paths.**

SUMMARY – RADIO WAVES, PROPAGATION

T3A09 When skip signals refracted from the ionosphere are elliptically polarized, **either vertically or horizontally polarized antennas may be used for transmission or reception.**

T3A10 If data signals arrive via multiple paths **error rates are likely to increase.**

T3A11 The part of the atmosphere that enables the propagation of radio signals around the world is the **ionosphere.**

T3A12 On 10 meters and 6 meters, **fog and light rain will have little effect.**

T3A13 Weather condition that decreases range at microwave frequencies is **precipitation.**

~~~~~~

T3B01 The distance a radio wave travels during one complete cycle is the **wavelength.**

T3B02 The property of a radio wave used to describe its polarization is the **orientation of the electric field.**

T3B03 The two components of a radio wave are the **electric and magnetic fields.**

T3B04 A radio wave travels through free space **at the speed of light.**

T3B05 The wavelength of a radio wave gets **shorter as the frequency increases.**

T3B06 The formula for converting frequency to approximate wavelength in meters is **300 divided by frequency in megahertz.** *Cheat: The only answer with "300 divided by."*

T3B07 The property of radio waves often used to identify the different frequency bands is **the approximate wavelength.**

SUMMARY – RADIO WAVES, PROPAGATION

T3B08 The frequency limits of the VHF spectrum are **30 to 300 MHz.**

T3B09 The frequency limits of the UHF spectrum are **300 to 3000 MHz.**

T3B10 The frequency range referred to as HF is **3 to 30 MHz.**

T3B11 The approximate velocity of a radio wave as it travels through free space is **300,000,000 meters per second.**

~~~~~~

T3C01 Direct (not via a repeater) UHF signals are rarely heard from stations outside your local coverage area because **UHF signals are usually not reflected by the ionosphere.**

T3C02 An advantage of HF vs VHF and higher frequencies is that **long distance ionospheric propagation is far more common on HF.**

T3C03 A characteristic of VHF signals received via auroral reflection is **the signals exhibit rapid fluctuations of strength and often sound distorted.**

T3C04 The propagation type most commonly associated with occasional strong over-the-horizon signals on the 10, 6, and 2 meter bands is **sporadic E.** *Cheat: If you see "sporadic E," that is correct.*

T3C05 Radio signals might be heard despite obstructions because of **knife-edge diffraction.**

T3C06 Over-the-horizon VHF and UHF communications to ranges of approximately 300 miles are due to **tropospheric scatter.** *Cheat: If you see "tropospheric scatter," that is correct.*

T3C07 The band is best suited for communicating via meteor scatter is **6 meters.**

T3C08 Tropospheric ducting is caused by **temperature inversions in the atmosphere.**

T3C09 The best time for long-distance 10 meter band propagation via the F layer is **from dawn to shortly after sunset during periods of high sunspot activity.** *Hint: Daylight and high sunspot activity.*

T3C10 The bands that may provide long distance communications during the peak of the sunspot cycle **are six or ten meters.** *Hint: HF Bands are long distance.*

T3C11 VHF and UHF radio signals usually travel somewhat farther than the visual line of sight distance between two stations because the **Earth seems less curved to radio waves than to light.**

SUBELEMENT T4 - Amateur radio practices and station set-up
[2 Exam Questions - 2 Groups]

T4A01 To determine the minimum current capacity needed for a transceiver's power supply consider:
Efficiency of the transmitter at full power output
Receiver and control circuit power
Power supply regulation and heat dissipation
All of these are correct.

T4A02 A computer can be used as part of an amateur radio station:
For logging contacts and contact information
For sending and/or receiving CW
For generating and decoding digital signals
All of these choices are correct.

T4A03 Wiring between the power source and radio should be heavy-gauge wire and kept as short as possible **to avoid voltage falling below that needed for proper operation.**

T4A04 The computer sound card port connected to a transceiver's headphone or speaker output for operating digital modes is the **microphone or line input.** *Hint: Output has to be connected to input.*

T4A05 The proper location for an external SWR meter is **in series with the feed line, between the transmitter and antenna.**

T4A06 The connections between a voice transceiver and a computer for digital operation are the **receive audio, transmit audio, and push-to-talk (PTT).**

T4A07 A **computer's sound card provides audio to the radio's microphone input and converts received audio to digital form** when conducting digital communications. *Hint: A sound card provides audio.*

T4A08 **Flat strap** provides the lowest impedance to RF signals.

T4A09 To cure distorted audio caused by RF current on the shield of a microphone cable use a **ferrite choke.**

T4A10 The source of a high-pitched whine that varies with engine speed in a mobile transceiver's receive audio is the **alternator.**

T4A11 The negative return connection of a mobile transceiver's power cable should be connected at the **battery or engine block ground strap.**

~~~~~~

T4B01 If a transmitter is operated with the microphone gain set too high, **the output signal might become distorted.**

T4B02 The operating frequency on a modern transceiver can be selected by the **keypad or VFO knob.**

T4B03 The purpose of the squelch control on a transceiver is to **mute receiver output noise when no signal is being received.**

T4B04 A way to enable quick access to a favorite frequency on your transceiver is to **store the frequency in a memory channel.**

T4B05 To reduce ignition interference to a receiver, **turn on the noise blanker.**

T4B06 If the voice pitch of a single-sideband signal seems too high or low **use the receiver RIT or clarifier.**

T4B07 The term "RIT" means **Receiver Incremental Tuning.**

T4B08 The advantage of having multiple receive bandwidth choices on a multimode transceiver it **permits noise or**

SUMMARY – PRACTICES

interference reduction by selecting a bandwidth matching the mode.

T4B09 An appropriate receive filter bandwidth for minimizing noise and interference for SSB reception is **2400 Hz.**

T4B10 An appropriate receive filter bandwidth for minimizing noise and interference for CW reception is **500 Hz.**

T4B11 The common meaning of the term "repeater offset" is the **difference between the repeater's transmit and receive frequencies.**

T4B12 The function of automatic gain control, or AGC is to **keep received audio relatively constant.**

T4B13 To remove power line noise or ignition noise, use the **noise blanker.**

T4B14 Use the scanning function of an FM transceiver to **scan through a range of frequencies to check for activity.**

SUBELEMENT T5 – Electrical principles: math for electronics; electronic principles; Ohm's Law

[4 Exam Questions - 4 Groups]

T5A01 Electrical current is measured in **Amperes.**

T5A02 Electrical power is measured in **Watts.**

T5A03 The name for the flow of electrons in an electric circuit is **Current.** *Hint: Current flows.*

T5A04 Current that flows only in one direction is **direct current.**

T5A05 The electrical term for the electromotive force (EMF) that causes electron flow is **voltage.**

T5A06 A mobile transceiver typically requires about **12 volts.** *Hint: Same as the car's electrical system.*

T5A07 **Copper** is a good electrical conductor

T5A08 **Glass** is a good electrical insulator.

T5A09 Current that reverses direction on a regular basis is **alternating current.**

T5A10 The rate at which electrical energy is used is **power.**

T5A11 The unit of electromotive force is the **volt.**

T5A12 The number of times per second that an alternating current makes a complete cycle is called **frequency.**

T5A13 The current is the same through all components in a **series** circuit.

T5A14 The voltage is the same across all components in a **parallel** circuit.

SUMMARY – ELECTRICAL PRINCIPLES

T5B01 1.5 amperes is **1500 milliamperes.**

T5B02 1,500,000 hertz is **1500 kHz.**

T5B03 One kilovolt is **one thousand** volts.

T5B04 One microvolt is **one-millionth** of a volt. *Hint: Micro is small – one millionth.*

T5B05 500 milliwatts is **0.5 watts.** *Hint: Milli is one-thousandth.*

T5B06 An ammeter calibrated in amperes used to measure a 3000-milliampere current would show **3 amperes.**

T5B07 A frequency display calibrated in megahertz reading 3.525 MHz, would show **3525 kHz** if it were calibrated in kilohertz.

T5B08 1,000,000 picofarads is the same as **1 microfarad.**

T5B09 A power increase from 5 watts to 10 watts is **3 dB.** *Hint: Double or half is 3 dB.*

T5B10 The change, measured in decibels (dB), of a power decrease from 12 watts to 3 watts is **-6 dB.** *Hint: 6 dB is one half and then one half again.*

T5B11 A power increase from 20 watts to 200 watts is **10 dB.** *Hint: Ten times is 10 dB.*

T5B12 28,400 kHz is the same as **28.400 MHz.**

T5B13 2425 MHz, is the same as **2.425 GHz.**

~~~~~~

T5C01 The ability to store energy in an electric field is called **capacitance.**

T5C02 The basic unit of capacitance is the **farad**

T5C03 The ability to store energy in a magnetic field is called **inductance.** *Remember: Electric field is capacitance, magnetic field is inductance.*

T5C04 The basic unit of inductance is the **henry.**

T5C05 The unit of frequency is **Hertz.**

T5C06 The abbreviation "RF" refers to **radio frequency signals of all types.**

T5C07 A radio wave is made up of **electromagnetic** energy.

T5C08 The formula used to calculate electrical power in a DC circuit is **power (P) equals voltage (E) multiplied by current (I).** *Cheat: Easy as PIE. P=IE.*

T5C09 The power used in a circuit when the applied voltage is 13.8 volts DC and the current is 10 amperes is **138 watts.** *Solve: 10 x 13.8 = 138.*

T5C10 The power used in a circuit when the applied voltage is 12 volts DC and the current is 2.5 amperes is **30 watts.** *Solve 12 x 2.5 = 30.*

T5C11 The amperes flowing in a circuit when the applied voltage is 12 volts DC and the load is 120 watts is **10 amperes.** *Solve: 120/12 = 10.*

T5C12 Impedance is a measure of the **opposition to AC current flow in a circuit.**

T5C13 The units of impedance are **ohms.** *Hint: Resistanceunits are the same as in a DC circuit.*

T5C14 The proper abbreviation for megahertz is **MHz.** *Remember capital M and capital H. Mr. Hertz.*

SUMMARY – ELECTRICAL PRINCIPLES

T5D01 The formula used to calculate current in a circuit is **current (I) equals voltage (E) divided by resistance (R.** *Hint: Use your magic circle, or you will get Confused*

T5D02 The formula used to calculate voltage in a circuit is **voltage (E) equals current (I) multiplied by resistance (R).**

T5D03 The formula used to calculate resistance in a circuit is **resistance (R) equals voltage (E) divided by current (I).**

T5D04 The resistance of a circuit in which a current of 3 amperes flows through a resistor connected to 90 volts is **30 ohms.** *Solve: R = E/I or 90 /3 =30*

T5D05 The resistance in a circuit for which the applied voltage is 12 volts and the current flow is 1.5 amperes is **8 ohms.** *Solve: R = E/I or 12 / 1.5 = 8.*

T5D06 The resistance of a circuit that draws 4 amperes from a 12-volt source is **3 ohms.** *Solve: R=E/I or 12/4 = 3.*

T5D07 The current in a circuit with an applied voltage of 120 volts and a resistance of 80 ohms is **1.5 amperes.** *Solve: No more hints, use your magic circle.*

T5D08 The current through a 100-ohm resistor connected across 200 volts is **2 amperes.**

T5D09 The current through a 24-ohm resistor connected across 240 volts is 10 **amperes**.

T5D10 The voltage across a 2-ohm resistor if a current of 0.5 amperes flows through it is **1 volt.**

T5D11 The voltage across a 10-ohm resistor if a current of 1 ampere flows through it is **10 volts.**

T5D12 The voltage across a 10-ohm resistor if a current of 2 amperes flows through it is **20 volts.**

T5D13 The current at the junction of two components in series is **unchanged**. *Remember: It is one path, so the same current still has to flow through.*

T5D14 The current at the junction of two components in parallel **divides between them** dependent on the value of the components. *Remember: In parallel, the current splits along the different paths.*

T5D15 The voltage across each of two components in series with a voltage source **is determined by the type and value of the components.** *Solve: The same current flows but the voltage (pressure) drops through each component in series.*

T5D16 The voltage across each of two components in parallel with a voltage source is **the same voltage as the source.** **Solve:** *In parallel, the pressure is the same, the current splits.*

SUBELEMENT T6 – Electrical components; circuit diagrams; component functions

[4 Exam Questions - 4 Groups]

T6A01 The electrical component that opposes the flow of current in a DC circuit is a **resistor**.

T6A02 The component often used as an adjustable volume control is a **potentiometer**.

T6A03 The electrical parameter controlled by a potentiometer is **resistance**. *Solve: It is a variable resistor.*

T6A04 The electrical component that stores energy in an electric field is a **capacitor.**

T6A05 The electrical component consisting of two or more conductive surfaces separated by an insulator is a **capacitor**.

T6A06 The electrical component that stores energy in a magnetic field is an **inductor.** *Remember: Electrical field, capacitor and magnetic field, inductor.*

T6A07 A coil of wire is an **inductor**

T6A08 The component used to connect or disconnect electrical circuits is a **switch**.

T6A09 The electrical component used to protect other circuit components from current overloads is a **fuse**.

T6A10 Rechargeable battery types are:
Nickel-metal hydride
Lithium-ion
Lead-acid gel-cell
All of these choices are correct.

T6A11 A **carbon-zinc** battery is not rechargeable.

SUMMARY – ELECTRICAL COMPONENTS, DIAGRAMS

T6B01 The electronic components that use a voltage or current signal to control current flow are **transistors.**

T6B02 The electronic component that allows current to flow in only one direction is a **diode.**

T6B03 The component that can be used as an electronic switch or amplifier is a **transistor.**

T6B04 A **transistor** can consist of three layers of semiconductor material.

T6B05 A **transistor** can amplify signals.

T6B06 The cathode lead of a semiconductor diode is often marked with a **stripe.**

T6B07 LED stands for **Light Emitting Diode.**

T6B08 The abbreviation FET stands for **Field Effect Transistor.**

T6B09 The names of the two electrodes of a diode are **anode and cathode.**

T6B10 The primary gain-producing component in an RF power amplifier is a **transistor.** *Remember: transistors can be switches or amplifiers.*

T6B11 The ability to amplify a signal is called **gain.**

T6C01 The name of an electrical wiring diagram that uses standard component symbols is a **schematic.**

Refer to the three diagrams at the end of the book (before the Index). You will get one diagram with your test.

T6C02 Component 1 in figure T1 is a **resistor.**

T6C03 Component 2 in figure T1 is a **transistor.**

SUMMARY – ELECTRICAL COMPONENTS, DIAGRAMS

T6C04 Component 3 in figure T1 is a **lamp.**

T6C05 Component 4 in figure T1 is a **battery**.

T6C06 Component 6 in figure T2 is a **capacitor**.

T6C07 Component 8 in figure T2 is a **light emitting diode**.

T6C08 Component 9 in figure T2 is a **variable resistor** (aka potentiometer). .

T6C09 Component 4 figure T2 is a **transformer**.

T6C10 Component 3 in figure T3 is a **variable inductor.**

T6C11 Component 4 in figure T3 is an **antenna.**

T6C12 The symbols on an electrical schematic represent **electrical components.**

T6C13 Electrical schematics accurately represented the **way components are interconnected.**

~~~~~~

T6D01 A **rectifier** changes an alternating current into a varying direct current signal.

T6D02 A **relay** is an electrically-controlled switch.

T6D03 The switch is represented by component 3 in figure T2 is **single-pole single-throw.**

T6D04 The component that displays an electrical quantity as a numeric value is a **meter.**

T6D05 The amount of voltage from a power supply is controlled by a **regulator.**

T6D06 The component commonly used to change 120V AC house current to a lower AC voltage for other uses is a

transformer. *Remember regulators regulate, transformers transform.*

T6D07 The component commonly used as a visual indicator is an **LED.**

T6D08 The component combined with an inductor to make a tuned circuit is a **capacitor.**

T6D09 The device that combines several semiconductors and other components into one package is an **integrated circuit.**

T6D10 The function of component 2 in Figure T1 is to **control the flow of current.** It is an example of a transistor used as a switch.

T6D11 A resonant or tuned circuit is **an inductor and a capacitor connected in series or parallel to form a filter.** *Remember a tuned circuit forms a filter.*

T6D12 A common reason to use shielded wire is **to prevent coupling of unwanted signals to or from the wire.**

SUBELEMENT T7 – Station equipment: common transmitter and receiver problems; antenna measurements; troubleshooting; basic repair and testing
[4 Exam Questions - 4 Groups]

T7A01 The term that describes the ability of a receiver to detect the presence of a signal is **sensitivity.**

T7A02 A transceiver is **a unit combining the functions of a transmitter and a receiver.**

T7A03 To convert a radio signal from one frequency to another, use a **mixer.**

T7A04 The term that describes the ability of a receiver to discriminate between multiple signals is **selectivity.**

T7A05 The circuit that generates a signal at a specific frequency is an **oscillator**

T7A06 The device that converts the RF input and output of a transceiver to another band is a **transverter.**

T7A07 "PTT" is the **push-to-talk function that switches between receive and transmit.**

T7A08 Combining speech with an RF carrier signal is called **modulation**.

T7A09 The function of the SSB/CW-FM switch on a VHF power amplifier is to **set the amplifier for proper operation in the selected mode.**

T7A10 The device that increases the low-power output from a handheld transceiver is an **RF power amplifier.**

T7A11 An RF preamplifier is installed **between the antenna and receiver.**

SUMMARY – EQUIPMENT, MEASUREMENTS

T7B01 If you are told your FM handheld or mobile transceiver is over-deviating **talk farther away from the microphone.**

T7B02 A broadcast AM or FM radio might receive an amateur radio transmission unintentionally because **the receiver is unable to reject strong signals outside the AM or FM band.**

T7B03 Radio frequency interference can be caused by:
Fundamental overload
Harmonics
Spurious emissions
All of these choices are correct

T7B04 A way to reduce or eliminate interference from an amateur transmitter to a nearby telephone is to **put an RF filter on the telephone.**

T7B05 Overload of a non-amateur radio or TV receiver by an amateur signal can be reduced or eliminated if you **block the amateur signal with a filter at the antenna input of the affected receiver.**

T7B06 If a neighbor tells you that your station's transmissions are interfering with their radio or TV reception, **make sure that your station is functioning properly and that it does not cause interference to your radio or television when it is tuned to the same channel.**

T7B07 To reduce overload to a VHF transceiver from a nearby FM broadcast station, use a **band-reject filter.**

T7B08 If something in a neighbor's home is causing harmful interference to your amateur station:
Work with your neighbor to identify the offending device
Politely inform your neighbor about the rules that prohibit the use of devices that cause interference
Check your station and make sure it meets the standards of good amateur practice
All of these choices are correct.

SUMMARY – EQUIPMENT, MEASUREMENTS

T7B09 A Part 15 device is an unlicensed device that **may emit low-powered radio signals on frequencies used by a licensed service.**

T7B10 If you receive a report that your audio signal through the repeater is distorted or unintelligible:
Your transmitter is slightly off frequency
Your batteries are running low
You are in a bad location
All of these choices are correct.

T7B11 A symptom of RF feedback in a transmitter or transceiver **is reports of garbled, distorted, or unintelligible voice transmissions.**

T7B12 The first step to resolve cable TV interference from your ham radio transmission is to **be sure all TV coaxial connectors are installed properly.**

T7C01 The primary purpose of a dummy load is to **prevent transmitting signals over the air when making tests.**

T7C02 To determine if an antenna is resonant at the desired operating frequency, use an **antenna analyzer.**

T7C03 The standing wave ratio (SWR) is a **measure of how well a load is matched to a transmission line.**

T7C04 The SWR meter reading indicating a perfect impedance match between the antenna and the feed line is **1 to 1.**

T7C05 Most solid-state amateur radio transmitters reduce output power as SWR increases **to protect the output amplifier transistors.**

T7C06 An SWR reading of 4:1 indicates an **impedance mismatch.**

SUMMARY – EQUIPMENT, MEASUREMENTS

T7C07 Power lost in a feed line **is converted into heat.**

T7C08 Other than an SWR meter, you could use **a directional wattmeter** to determine if a feed line and antenna are properly matched.

T7C09 The most common cause for failure of coaxial cables is **moisture contamination.**

T7C10 The outer jacket of coaxial cable should be resistant to ultraviolet light because **ultraviolet light can damage the jacket and allow water to enter the cable.**

T7C11 The disadvantage of air core coaxial cable when compared to foam or solid dielectric types is **it requires special techniques to prevent water absorption.**
Remember: "water" is in all three answers.

T7C12 A dummy load consists of a **non-inductive resistor and a heat sink.**

T7D01 To measure electric potential or electromotive force, use a **voltmeter.**

T7D02 The correct way to connect a voltmeter to a circuit is **in parallel with the circuit.** (Across the circuit).

T7D03 An ammeter is connected **in series with the circuit.** (Measure flow inline)

T7D04 To measure electric current, use an **ammeter.**

T7D05 To measure resistance, use an **ohmmeter.**

T7D06 You might damage a multimeter attempting to measure voltage when **using the resistance setting.**

T7D07 The measurements commonly made using a multimeter are **voltage and resistance.**

SUMMARY – EQUIPMENT, MEASUREMENTS

T7D08 The solder best for radio and electronic use is **rosin-core solder.**

T7D09 A cold solder joint has a **grainy or dull surface.**

T7D10 When an ohmmeter, connected across an unpowered circuit, initially indicates a low resistance and then shows increasing resistance with time, **the circuit contains a large capacitor.**

T7D11 When measuring circuit resistance with an ohmmeter, **ensure that the circuit is not powered.**

T7D12 When measuring high voltages with a voltmeter, **ensure that the voltmeter and leads are rated for use at the voltages to be measured.**

SUBELEMENT T8 – Modulation modes: amateur satellite operation; operating activities; non-voice and digital communications
[4 Exam Questions - 4 Groups]

T8A01 **Single sideband** is a form of amplitude modulation.

T8A02 The modulation most commonly used for VHF packet radio transmissions is **FM.**

T8A03 The voice mode most often used for long-distance (weak signal) contacts on the VHF and UHF bands is **SSB.**

T8A04 The type of modulation most commonly used for VHF and UHF voice repeaters is **FM**.

T8A05 The type of emission with the narrowest bandwidth is **CW.**

T8A06 The sideband normally used for 10 meter HF, VHF, and UHF single-sideband communications is **upper sideband.**

T8A07 An advantage of single sideband (SSB) over FM for voice transmissions is **SSB signals have narrower bandwidth.**

T8A08 The approximate bandwidth of a single sideband (SSB) voice signal is **3 kHz.**

T8A09 The approximate bandwidth of a VHF repeater FM phone signal is **between 10 and 15 kHz.**

T8A10 The typical bandwidth of analog fast-scan TV transmissions on the 70 centimeter band is **about 6 MHz.**

T8A11 The approximate maximum bandwidth required to transmit a CW signal is **150 Hz.**

SUMMARY – MODULATION, SATELLITES, DIGITAL

T8B01 Telemetry information typically transmitted by satellite beacons is **health and status of the satellite.**

T8B02 The impact of using too much effective radiated power on a satellite uplink is **blocking access by other users.**

T8B03 Satellite tracking programs provide:
Maps showing the real-time position of the satellite track over the earth
The time, azimuth, and elevation of the start, maximum altitude, and end of a pass
The apparent frequency of the satellite transmission, including effects of Doppler shift
All of these choices are correct
Hint: All the things you need to know to track a satellite.

T8B04 The mode of transmission used for satellite beacons is:
RTTY
CW
Packet
All of these choices are correct.

T8B05 A satellite beacon is a transmission from a satellite that **contains status information.**

T8B06 The inputs to a satellite tracking program are the **Keplerian elements.**

T8B07 Doppler shift is an **observed change in signal frequency caused by relative motion between the satellite and the earth station.**

T8B08 If a satellite is operating in mode U/V, **the satellite uplink is in the 70 centimeter band and the downlink is in the 2 meter band.**

T8B09 Spin fading of satellite signals is caused by **rotation of the satellite and its antennas.**

T8B10 The initials LEO tell you the satellite is in a **Low Earth Orbit.**

T8B11 **Anyone who can receive the telemetry signal** may receive telemetry from a space station. *Hint: There are no restrictions on receiving, just transmitting.*

T8B12 A good way to judge whether your uplink power is neither too low nor too high is **your signal strength on the downlink should be about the same as the beacon.**

T8C01 A method used to locate sources of noise interference or jamming **is radio direction finding.**

T8C02 A useful item for a hidden transmitter hunt is a **directional antenna.**

T8C03 The operating activity involving contacting as many stations as possible during a specified period is called **contesting.**

T8C04 A good procedure when contacting another station in a radio contest is to **send only the minimum information needed for proper identification and the contest exchange.**

T8C05 A grid locator is **a letter-number designator assigned to a geographic location.**

T8C06 Access to some IRLP nodes is accomplished **by using DTMF signals.**

T8C07 Voice Over Internet Protocol (VoIP) is a **method of delivering voice communications over the internet using digital techniques.**

T8C08 The Internet Radio Linking Project (IRLP) is a **technique to connect amateur radio systems, such as repeaters, via the internet using Voice Over Internet Protocol** (VOIP). *Remember IRLP uses VoIP.*

SUMMARY – MODULATION, SATELLITES, DIGITAL

T8C09 You can obtain a list of active nodes that use VoIP:
By subscribing to an on line service
From on line repeater lists maintained by the local repeater frequency coordinator
From a repeater directory
All of these choices are correct.

T8C10 Before you may use the Echolink system to communicate using a repeater, **you must register your call sign and provide proof of license.**

T8C11 The name given to an amateur radio station that is used to connect other amateur stations to the internet is a **gateway.**

~~~~~~

T8D01 The following are digital communications modes:
Packet radio
IEEE 802.11
JT65
All of these choices are correct.

T8D02 The term "APRS" means **Automatic Packet Reporting System.**

T8D03 The device used to provide data to the transmitter when sending automatic position reports from a mobile amateur radio station is a **Global Positioning System receiver.**

T8D04 The type of transmission indicated by the term "NTSC" is an **analog fast scan color TV signal.**

T8D05 An application of APRS (Automatic Packet Reporting System) is **providing real-time tactical digital communications in conjunction with a map showing the locations of stations.** *Remember: Location of stations.*

T8D06 The abbreviation "PSK" means **Phase Shift Keying.**

T8D07 DMR (Digital Mobile Radio or Digital Migration Radio) is a **technique for time-multiplexing two digital voice** signals on a single 12.5 kHz repeater channel. *Remember: "time multiplexing."*

T8D08 Included in packet transmissions are:
A check sum that permits error detection
A header that contains the call sign of the station to which the information is being sent
Automatic repeat request in case of error
All of these choices are correct.

T8D09 The code used when sending CW in the amateur bands is **International Morse.**

T8D10 Digital mode software in the WSJT suite supports:
Moonbounce or Earth-Moon-Earth
Weak-signal propagation beacons
Meteor scatter
All of these choices are correct.

T8D11 An ARQ transmission system is a digital scheme whereby the receiving station **detects errors and sends a request to the sending station to retransmit the information.**

T8D12 Broadband-Hamnet(TM), also referred to as **a high-speed multi-media network is an amateur-radio-based data network using commercial Wi-Fi gear with modified firmware.** *Remember: It uses commercial Wi-Fi gear.*

T8D13 FT8 is a digital mode capable of operating in **low signal-to-noise conditions that transmits on 15-second intervals.**

T8D14 An electronic keyer is a device that **assists in manual sending of Morse code.**

SUBELEMENT T9 – Antennas and feed lines
[2 Exam Questions - 2 Groups]

T9A01 A beam antenna is an antenna that **concentrates signals in one direction.**

T9A02 One type of antenna loading is **inserting an inductor in the radiating portion of the antenna to make it electrically longer.**

T9A03 A simple dipole oriented parallel to the Earth's surface is **a horizontally polarized antenna.**

T9A04 A disadvantage of the "rubber duck" antenna supplied with most handheld radio transceivers when compared to a full-sized quarter-wave antenna is **it does not transmit or receive as effectively.**

T9A05 To change a dipole antenna to make it resonant on a higher frequency, **shorten it.**

T9A06 The quad, Yagi, and dish are **directional antennas.**

T9A07 A disadvantage of using a handheld VHF transceiver, with its integral antenna, inside a vehicle is **signals might not propagate well due to the shielding effect of the vehicle.**

T9A08 The approximate length, in inches, of a quarter-wavelength vertical antenna for 146 MHz is **19 inches.** *Solve: 146 MHz is 2 meters. A full wave is 2 meters long so 1/4 wave is 1/2 meter or about 19 inches.*

T9A09 The approximate length, in inches, of a half-wavelength 6 meter dipole antenna is **112 inches.** *Solve: A full wave is 6 meters, so a half wave is 3 meters, or about 3 x 39 inches. This is the closest answer.*

T9A10 A half-wave dipole antenna radiate its strongest signal **broadside to the antenna.**

T9A11 The gain of an antenna is the **increase in signal strength in a specified direction** compared to a reference antenna.

T9A12 An advantage of using a properly mounted 5/8 wavelength antenna for VHF or UHF mobile service is that **it has a lower radiation angle and more gain than a 1/4 wavelength antenna.**

~~~~~~

T9B01 It is important to have low SWR when using coaxial cable feed line **to reduce signal loss.**

T9B02 The impedance of most coaxial cables used in amateur radio installations is **50 ohms.**

T9B03 Coaxial cable is the most common feed line selected for amateur radio antenna systems because **it is easy to use and requires few special installation considerations.** *Remember: It is more convenient.*

T9B04 The major function of an antenna tuner (antenna coupler) **is it matches the antenna system impedance to the transceiver's output impedance.**

T9B05 As the frequency of a signal passing through coaxial cable is increased, **the loss increases.** *Remember: Higher frequencies, higher losses.*

T9B06 The connector most suitable for frequencies above 400 MHz is a **Type N connector.**

T9B07 PL-259 type coax connectors are **commonly used at HF frequencies.**

T9B08 Coax connectors exposed to the weather should be sealed against water intrusion **to prevent an increase in feed line loss.**

SUMMARY – ANTENNAS, FEEDLINES

T9B09 Erratic changes in SWR readings can be caused by **a loose connection in an antenna or a feed line.**

T9B10 The electrical difference between RG-58 and RG-8 coaxial cable is **RG-8 cable has less loss at a given frequency.**

T9B11 The feed line with the lowest loss at VHF and UHF is **air-insulated hard line.**

SUBELEMENT T0 – Electrical safety: AC and DC power circuits; antenna installation; RF hazards

[3 Exam Questions - 3 Groups]

T0A01 A safety hazard of a 12-volt storage battery **shorting the terminals can cause burns, fire, or an explosion.**

T0A02 The health hazard presented by electrical current flowing through the body is
It may cause injury by heating tissue
It may disrupt the electrical functions of cells
It may cause involuntary muscle contractions
All of these choices are correct.

T0A03 The green wire in a three-wire electrical AC plug is connected to the **equipment ground.**

T0A04 The purpose of a fuse in an electrical circuit is **to interrupt power in case of overload.**

T0A05 It is unwise to install a 20-ampere fuse in the place of a 5-ampere fuse because **excessive current could cause a fire.**

T0A06 To guard against electrical shock at your station:
Use three-wire cords and plugs
Connect all AC powered equipment to a common safety ground
Use a circuit protected by a ground-fault interrupter
All of these choices are correct.

T0A07 A precaution that should be taken when installing devices for lightning protection in a coaxial cable feed line is to **mount all of the protectors on a metal plate that is in turn connected to an external ground rod.**

T0A08 The safety equipment that should always be included in home-built equipment that is powered from 120V AC power circuits is a **fuse or circuit breaker in series with the AC hot conductor.**

SUMMARY – SAFETY

T0A09 All external ground rods or earth connections should be **bonded together with heavy wire or conductive strap.**

T0A10 If a lead-acid storage battery is charged or discharged too quickly, **the battery could overheat, give off flammable gas, or explode.**

T0A11 A hazard that might exist in a power supply when it is turned off and disconnected is **you might receive an electric shock from the charge stored in large capacitors.**

T0B01 Members of a tower work team should wear a hard hat and safety glasses **at all times when any work is being done on the tower.**

T0B02 A good precaution to observe before climbing an antenna tower is **put on a carefully inspected climbing harness (fall arrester) and safety glasses.**

T0B03 When is it safe to climb a tower without a helper or observer? **Never.**

T0B04 An important safety precaution to observe when putting up an antenna tower is **to look for and stay clear of any overhead electrical wires.**

T0B05 The purpose of a gin pole is **to lift tower sections or antennas.**

T0B06 The minimum safe distance from a power line to allow when installing an antenna is **enough so that if the antenna falls unexpectedly, no part of it can come closer than 10 feet to the power wires.**

T0B07 An important safety rule to remember when using a crank-up tower is **this type of tower must not be climbed unless retracted or mechanical safety locking devices have been installed.**

T0B08 A proper grounding method for a tower is a **separate eight-foot long ground rod for each tower leg, bonded to the tower and each other.**

T0B09 You should avoid attaching an antenna to a utility pole because **the antenna could contact high-voltage power lines.**

T0B10 When installing grounding conductors used for lightning protection, **sharp bends must be avoided.**

T0B11 Grounding requirements for an amateur radio tower or antenna are established by **local electrical codes.**

T0B12 When installing ground wires on a tower for lightning protection, **ensure that connections are short and direct.**

T0B13 The purpose of a safety wire through a turnbuckle used to tension guy lines is **to prevent loosening of the guy line from vibration.**

T0C01 VHF and UHF radio signals are **non-ionizing radiation.**

T0C02 The lowest value for Maximum Permissible Exposure limit is on **50 MHz.**

T0C03 The maximum power level that an amateur radio station may use at VHF frequencies before an RF exposure evaluation is required is **50 watts PEP at the antenna.** *Remember "at the antenna."*

T0C04 The factors which affect the RF exposure of people near an amateur station antenna are:
Frequency and power level of the RF field
Distance from the antenna to a person
Radiation pattern of the antenna
All of these choices are correct.

SUMMARY – SAFETY

T0C05 Exposure limits vary with frequency because **the human body absorbs more RF energy at some frequencies than at others.**

T0C06 The following is an acceptable method to determine that your station complies with FCC RF exposure regulations: **By calculation based on FCC OET Bulletin 65 By calculation based on computer modeling By measurement of field strength using calibrated equipment All of these choices are correct.**

T0C07 If a person accidentally touched your antenna while you were transmitting, **they might receive a painful RF burn.**

T0C08 To prevent exposure to RF radiation in excess of FCC-supplied limits **relocate antennas.**

T0C09 You can make sure your station stays in compliance with RF safety regulations **by re-evaluating the station whenever an item of equipment is changed.**

T0C10 Duty cycle is one of the factors used to determine safe RF radiation exposure levels **because it affects the average exposure of people to radiation.**

T0C11 The definition of duty cycle during the averaging time for RF exposure is **the percentage of time that a transmitter is transmitting.**

T0C12 RF radiation differs from ionizing radiation (radioactivity) because RF radiation **does not have sufficient energy to cause genetic damage.**

T0C13 If the averaging time for exposure is 6 minutes, how much power density is permitted if the signal is present for 3 minutes and absent for 3 minutes rather than being present for the entire 6 minutes? **Two times as much.**

END OF QUESTIONS

DIAGRAMS REQUIRED FOR EXAMINATIONS

You will be given one diagram with your test booklet.

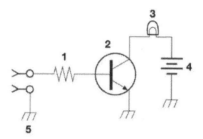

Figure T-1

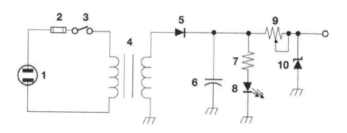

Figure T-2

DIAGRAMS

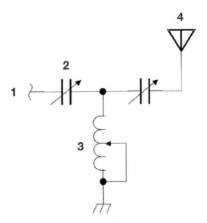

Figure T-3

Technician Class – The Easy Way

INDEX

INDEX

Made in the USA
San Bernardino, CA
23 July 2020

75885284R00084